BRITANNICA

Mathematics in Context

Growth

Britannica

ENCYCLOPÆDIA BRITANNICA EDUCATIONAL CORPORATION

Mathematics in Context is a comprehensive curriculum for the middle grades. It was developed in collaboration with the Wisconsin Center for Education Research, School of Education, University of Wisconsin–Madison and the Freudenthal Institute at the University of Utrecht, The Netherlands, with the support of National Science Foundation Grant No. 9054928.

National Science Foundation

Opinions expressed are those of the authors
and not necessarily those of the Foundation

ISBN 0-7826-1550-3
1 2 3 4 5 WK 02 01 00 99 98

The *Mathematics in Context* Development Team

Mathematics in Context is a comprehensive curriculum for the middle grades. The National Science Foundation funded the National Center for Research in Mathematical Sciences Education at the University of Wisconsin–Madison to develop and field-test the materials from 1991 through 1996. The Freudenthal Institute at the University of Utrecht in The Netherlands, as a subcontractor, collaborated with the University of Wisconsin–Madison on the development of the curriculum.

The initial version of *Growth* was developed by Anton Roodhardt. It was adapted for use in American schools by Mary S. Spence, Jack Burrill, and Peter Christiansen.

National Center for Research in Mathematical Sciences Education Staff

Thomas A. Romberg
Director

Gail Burrill
Coordinator
Field Test Materials

Mary Ann Fix
Editorial Coordinator

James A. Middleton
Pilot Test Coordinator

Joan Daniels Pedro
Assistant to the Director

Margaret R. Meyer
Coordinator
Pilot Test Materials

Sherian Foster
Editorial Coordinator

Margaret A. Pligge
First Edition Coordinator

Project Staff

Jonathan Brendefur
Laura J. Brinker
James Browne
Jack Burrill
Rose Byrd
Peter Christiansen
Barbara Clarke
Doug Clarke
Beth R. Cole
Fae Dremock
Jasmina Milinkovic
Kay Schultz
Mary C. Shafer
Julia A. Shew
Aaron N. Simon
Marvin Smith
Stephanie Z. Smith
Mary S. Spence
Kathleen A. Steele

Freudenthal Institute Staff

Jan de Lange
Director

Els Feijs
Coordinator

Martin van Reeuwijk
Coordinator

Project Staff

Mieke Abels
Nina Boswinkel
Frans van Galen
Koeno Gravemeijer
Marja van den Heuvel-Panhuizen
Jan Auke de Jong
Vincent Jonker
Ronald Keijzer
Martin Kindt
Jansie Niehaus
Nanda Querelle
Anton Roodhardt
Leen Streefland
Adri Treffers
Monica Wijers
Astrid de Wild

Acknowledgments

Several school districts used and evaluated one or more versions of the materials: Ames Community School District, Ames, Iowa; Parkway School District, Chesterfield, Missouri; Stoughton Area School District, Stoughton, Wisconsin; Madison Metropolitan School District, Madison, Wisconsin; Milwaukee Public Schools, Milwaukee, Wisconsin; and Dodgeville School District, Dodgeville, Wisconsin. Two sites were involved in staff development as well as formative evaluation of materials: Culver City, California, and Memphis, Tennessee. Two sites were developed through partnership with Encyclopædia Britannica Educational Corporation: Miami, Florida, and Puerto Rico. University Partnerships were developed with mathematics educators who worked with preservice teachers to familiarize them with the curriculum and to obtain their advice on the curriculum materials. The materials were also used at several other schools throughout the United States.

We at Encyclopædia Britannica Educational Corporation extend our thanks to all who had a part in making this program a success. Some of the participants instrumental in the program's development are as follows:

Allapattah Middle School
Miami, Florida
Nemtalla (Nikolai) Barakat

Ames Middle School
Ames, Iowa
Kathleen Coe
Judd Freeman
Gary W. Schnieder
Ronald H. Stromen
Lyn Terrill

Bellerive Elementary
Creve Coeur, Missouri
Judy Hetterscheidt
Donna Lohman
Gary Alan Nunn
Jakke Tchang

Brookline Public Schools
Brookline, Massachusetts
Rhonda K. Weinstein
Deborah Winkler

Cass Middle School
Milwaukee, Wisconsin
Tami Molenda
Kyle F. Witty

Central Middle School
Waukesha, Wisconsin
Nancy Reese

Craigmont Middle School
Memphis, Tennessee
Sharon G. Ritz
Mardest K. VanHooks

Crestwood Elementary
Madison, Wisconsin
Diane Hein
John Kalson

Culver City Middle School
Culver City, California
Marilyn Culbertson
Joel Evans
Joy Ellen Kitzmiller
Patricia R. O'Connor
Myrna Ann Perks, Ph.D.
David H. Sanchez
John Tobias
Kelley Wilcox

Cutler Ridge Middle School
Miami, Florida
Lorraine A. Valladares

Dodgeville Middle School
Dodgeville, Wisconsin
Jacqueline A. Kamps
Carol Wolf

Edwards Elementary
Ames, Iowa
Diana Schmidt

Fox Prairie Elementary
Stoughton, Wisconsin
Tony Hjelle

Grahamwood Elementary
Memphis, Tennessee
M. Lynn McGoff
Alberta Sullivan

Henry M. Flagler Elementary
Miami, Florida
Frances R. Harmon

Horning Middle School
Waukesha, Wisconsin
Connie J. Marose
Thomas F. Clark

Huegel Elementary
Madison, Wisconsin
Nancy Brill
Teri Hedges
Carol Murphy

Hutchison Middle School
Memphis, Tennessee
Maria M. Burke
Vicki Fisher
Nancy D. Robinson

Idlewild Elementary
Memphis, Tennessee
Linda Eller

Jefferson Elementary
Santa Ana, California
Lydia Romero-Cruz

Jefferson Middle School
Madison, Wisconsin
Jane A. Beebe
Catherine Buege
Linda Grimmer
John Grueneberg
Nancy Howard
Annette Porter
Stephen H. Sprague
Dan Takkunen
Michael J. Vena

Jesus Sanabria Cruz School
Yabucoa, Puerto Rico
Andreíta Santiago Serrano

John Muir Elementary School
Madison, Wisconsin
Julie D'Onofrio
Jane M. Allen-Jauch
Kent Wells

Kegonsa Elementary
Stoughton, Wisconsin
Mary Buchholz
Louisa Havlik
Joan Olsen
Dominic Weisse

Linwood Howe Elementary
Culver City, California
Sandra Checel
Ellen Thireos

Mitchell Elementary
Ames, Iowa
Henry Gray
Matt Ludwig

New School of Northern Virginia
Fairfax, Virginia
Denise Jones

Northwood Elementary
Ames, Iowa
Eleanor M. Thomas

Orchard Ridge Elementary
Madison, Wisconsin
Mary Paquette
Carrie Valentine

Parkway West Middle School
Chesterfield, Missouri
Elissa Aiken
Ann Brenner
Gail R. Smith

Ridgeway Elementary
Ridgeway, Wisconsin
Lois Powell
Florence M. Wasley

Roosevelt Elementary
Ames, Iowa
Linda A. Carver

Roosevelt Middle
Milwaukee, Wisconsin
Sandra Simmons

Ross Elementary
Creve Coeur, Missouri
Annette Isselhard
Sheldon B. Korklan
Victoria Linn
Kathy Stamer

St. Joseph's School
Dodgeville, Wisconsin
Rita Van Dyck
Sharon Wimer

St. Maarten Academy
St. Peters, St. Maarten, NA
Shareed Hussain

Sarah Scott Middle School
Milwaukee, Wisconsin
Kevin Haddon

Sawyer Elementary
Ames, Iowa
Karen Bush Hoiberg

Sennett Middle School
Madison, Wisconsin
Brenda Abitz
Lois Bell
Shawn M. Jacobs

Sholes Middle School
Milwaukee, Wisconsin
Chris Gardner
Ken Haddon

Stephens Elementary
Madison, Wisconsin
Katherine Hogan
Shirley M. Steinbach
Kathleen H. Vegter

Stoughton Middle School
Stoughton, Wisconsin
Sally Bertelson
Polly Goepfert
Jacqueline M. Harris
Penny Vodak

Toki Middle School
Madison, Wisconsin
Gail J. Anderson
Vicky Grice
Mary M. Ihlenfeldt
Steve Jernegan
Jim Leidel
Theresa Loehr
Maryann Stephenson
Barbara Takkunen
Carol Welsch

Trowbridge Elementary
Milwaukee, Wisconsin
Jacqueline A. Nowak

W. R. Thomas Middle School
Miami, Florida
Michael Paloger

Wooddale Elementary Middle School
Memphis, Tennessee
Velma Quinn Hodges
Jacqueline Marie Hunt

Yahara Elementary
Stoughton, Wisconsin
Mary Bennett
Kevin Wright

Site Coordinators

Mary L. Delagardelle—Ames Community Schools, Ames, Iowa

Dr. Hector Hirigoyen—Miami, Florida

Audrey Jackson—Parkway School District, Chesterfield, Missouri

Jorge M. López—Puerto Rico

Susan Militello—Memphis, Tennessee

Carol Pudlin—Culver City, California

Reviewers and Consultants

Michael N. Bleicher
Professor of Mathematics
University of Wisconsin–Madison
Madison, WI

Diane J. Briars
Mathematics Specialist
Pittsburgh Public Schools
Pittsburgh, PA

Donald Chambers
Director of Dissemination
University of Wisconsin–Madison
Madison, WI

Don W. Collins
Assistant Professor of Mathematics Education
Western Kentucky University
Bowling Green, KY

Joan Elder
Mathematics Consultant
Los Angeles Unified School District
Los Angeles, CA

Elizabeth Fennema
Professor of Curriculum and Instruction
University of Wisconsin-Madison
Madison, WI

Nancy N. Gates
University of Memphis
Memphis, TN

Jane Donnelly Gawronski
Superintendent
Escondido Union High School
Escondido, CA

M. Elizabeth Graue
Assistant Professor of Curriculum and Instruction
University of Wisconsin–Madison
Madison, WI

Jodean E. Grunow
Consultant
Wisconsin Department of Public Instruction
Madison, WI

John G. Harvey
Professor of Mathematics and Curriculum & Instruction
University of Wisconsin–Madison
Madison, WI

Simon Hellerstein
Professor of Mathematics
University of Wisconsin–Madison
Madison, WI

Elaine J. Hutchinson
Senior Lecturer
University of Wisconsin–Stevens Point
Stevens Point, WI

Richard A. Johnson
Professor of Statistics
University of Wisconsin–Madison
Madison, WI

James J. Kaput
Professor of Mathematics
University of Massachusetts–Dartmouth
Dartmouth, MA

Richard Lehrer
Professor of Educational Psychology
University of Wisconsin–Madison
Madison, WI

Richard Lesh
Professor of Mathematics
University of Massachusetts–Dartmouth
Dartmouth, MA

Mary M. Lindquist
Callaway Professor of Mathematics Education
Columbus College
Columbus, GA

Baudilio (Bob) Mora
Coordinator of Mathematics & Instructional Technology
Carrollton-Farmers Branch Independent School District
Carrollton, TX

Paul Trafton
Professor of Mathematics
University of Northern Iowa
Cedar Falls, IA

Norman L. Webb
Research Scientist
University of Wisconsin–Madison
Madison, WI

Paul H. Williams
Professor of Plant Pathology
University of Wisconsin–Madison
Madison, WI

Linda Dager Wilson
Assistant Professor
University of Delaware
Newark, DE

Robert L. Wilson
Professor of Mathematics
University of Wisconsin–Madison
Madison, WI

TABLE OF CONTENTS

Dear Teacher,

Welcome! *Mathematics in Context* is designed to reflect the National Council of Teachers of Mathematics Standards for School Mathematics and to ground mathematical content in a variety of real-world contexts. Rather than relying on you to explain and demonstrate generalized definitions, rules, or algorithms, students investigate questions directly related to a particular context and construct mathematical understanding and meaning from that context.

The curriculum encompasses 10 units per grade level. *Growth* is designed to be one of the four grade 8/9 algebra strand units. This unit builds on what students have learned in the grade 7/8 units *Building Formulas* and *Ups and Downs* and the grade 8/9 unit *Graphing Equations*. The unit also lends itself to independent use—to introduce students to exponential growth, recursive and direct formulas, and mathematical modeling.

In addition to the Teacher Guide and Student Books, *Mathematics in Context* offers the following components that will inform and support your teaching:

- *Teacher Resource and Implementation Guide,* which provides an overview of the complete system, including program implementation, philosophy, and rationale
- *Number Tools,* Volumes 1 and 2, which are a series of blackline masters that serve as review sheets or practice pages involving number issues and basic skills
- *News in Numbers,* which is a set of additional activities that can be inserted between or within other units; it includes a number of measurement problems that require estimation.

Thank you for choosing *Mathematics in Context*. We wish you success and inspiration!

Sincerely,

The Mathematics in Context Development Team

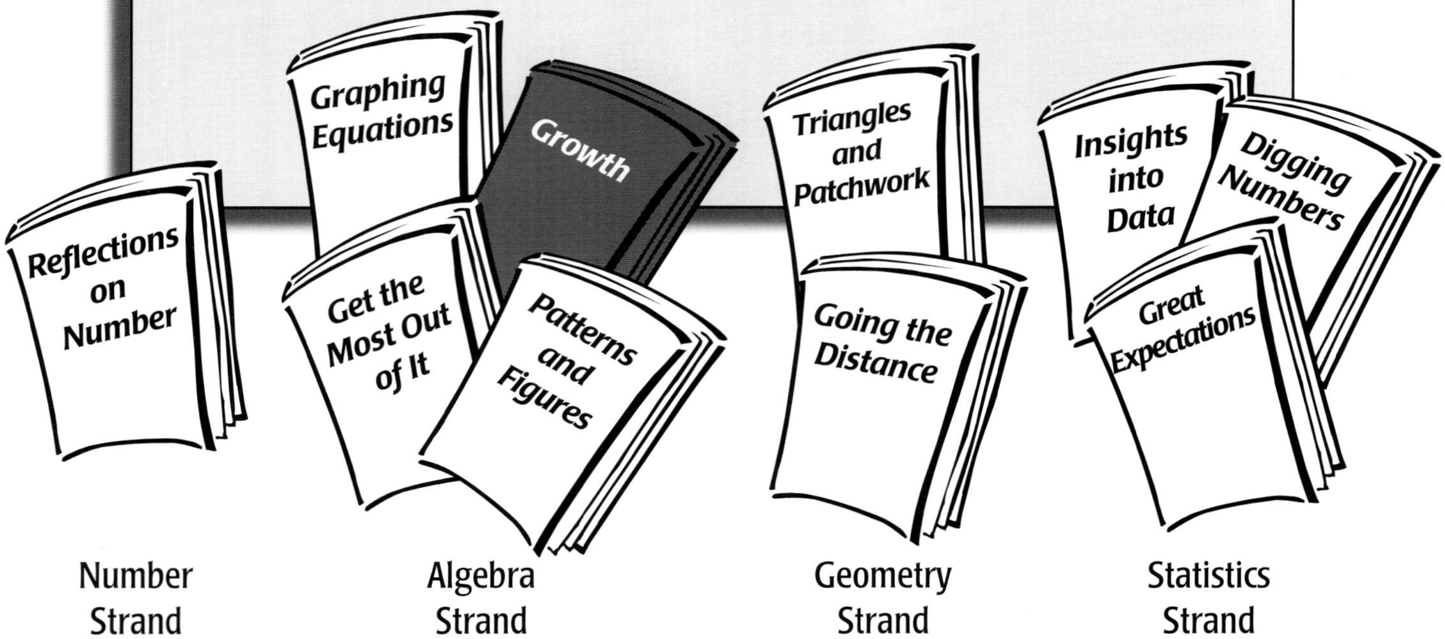

Overview

BRITANNICA
Mathematics in Context
BUS SCHEDULE
OUTDOOR
96.8 °F
77.5 °F
INDOOR
Gino's
W
N
S
E

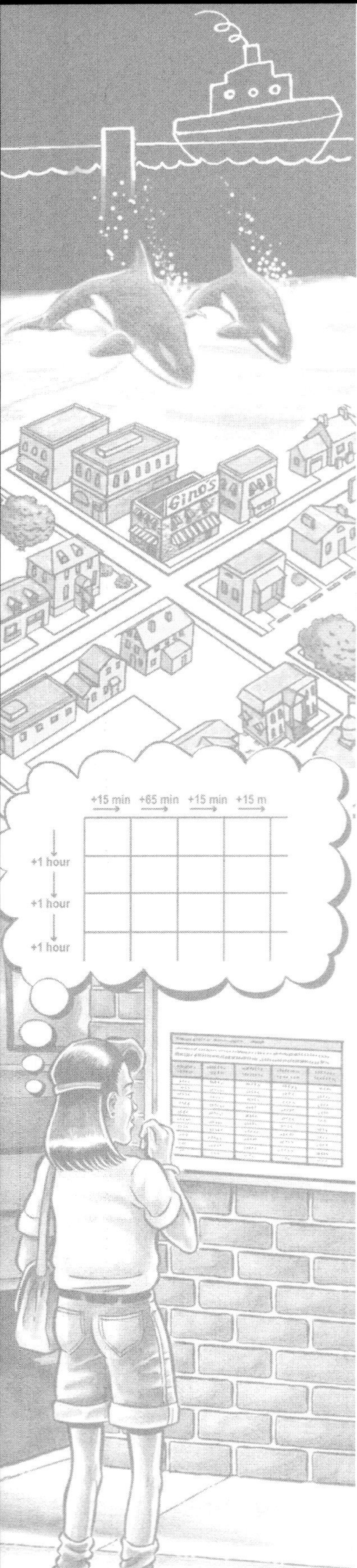

How to Use This Book

This unit is one of 40 for the middle grades. Each unit can be used independently; however, the 40 units are designed to make up a complete, connected curriculum (10 units per grade level). There is a Student Book and a Teacher Guide for each unit.

Each Teacher Guide comprises elements that assist the teacher in the presentation of concepts and in understanding the general direction of the unit and the program as a whole. Becoming familiar with this structure will make using the units easier.

Each Teacher Guide consists of six basic parts:

- Overview
- Student Materials and Teaching Notes
- Assessment Activities and Solutions
- Glossary
- Blackline Masters
- Try This! Solutions

Overview

Before beginning this unit, read the Overview in order to understand the purpose of the unit and to develop strategies for facilitating instruction. The Overview provides helpful information about the unit's focus, pacing, goals, and assessment, as well as explanations about how the unit fits with the rest of the *Mathematics in Context* curriculum.

Student Material and Teaching Notes

This Teacher Guide contains all of the student pages, (except the Try This! activities), each of which faces a page of solutions, samples of students' work, and hints and comments about how to facilitate instruction. Note: Solutions for the Try This! activities can be found at the back of the Teacher Guide.

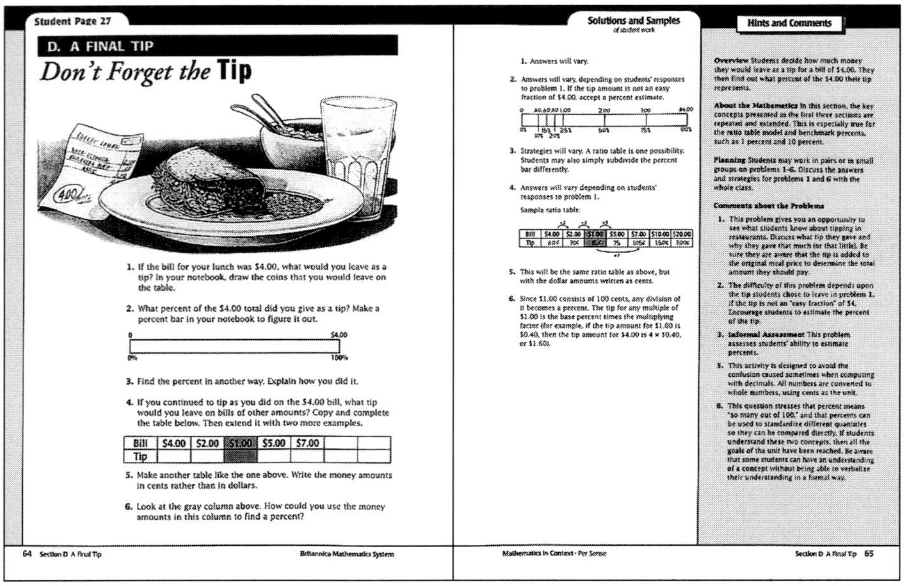

Student Page 27

D. A FINAL TIP

Don't Forget the **Tip**

1. If the bill for your lunch was $4.00, what would you leave as a tip? In your notebook, draw the coins that you would leave on the table.
2. What percent of the $4.00 total did you give as a tip? Make a percent bar in your notebook to figure it out.

0 $4.00
0% 100%

3. Find the percent in another way. Explain how you did it.
4. If you continued to tip as you did on the $4.00 bill, what tip would you leave on bills of other amounts? Copy and complete the table below. Then extend it with two more examples.

Bill	$4.00	$2.00	$1.00	$5.00	$7.00		
Tip							

5. Make another table like the one above. Write the money amounts in cents rather than in dollars.
6. Look at the gray column above. How could you use the money amounts in this column to find a percent?

64 Section D A Final Tip — Britannica Mathematics System

Solutions and Samples of student work

1. Answers will vary.
2. Answers will vary, depending on students' responses to problem 1. If the tip amount is not an easy fraction of $4.00, accept a percent estimate.
3. Strategies will vary. A ratio table is one possibility. Students may also simply subdivide the percent bar differently.
4. Answers will vary depending on students' responses to problem 1.

Sample ratio table:

Bill	$4.00	$2.00	$1.00	$5.00	$7.00	$10.00	$20.00
Tip	60¢	30¢	15¢	75	105¢	150¢	300¢

5. This will be the same ratio table as above, but with the dollar amounts written as cents.
6. Since $1.00 consists of 100 cents, any division of it becomes a percent. The tip for any multiple of $1.00 is the base percent times the multiplying factor (for example, if the tip amount for $1.00 is $0.40, then the tip amount for $4.00 is 4 × $0.40, or $1.60).

Mathematics in Context · Per Sense

Hints and Comments

Overview Students decide how much money they would leave as a tip for a bill of $4.00. They then find out what percent of the $4.00 their tip represents.

About the Mathematics In this section, the key concepts presented in the first three sections are repeated and extended. This is especially true for the ratio table model and benchmark percents, such as 1 percent and 10 percent.

Planning Students may work in pairs or in small groups on problems 1–6. Discuss the answers and strategies for problems 1 and 6 with the whole class.

Comments about the Problems

1. This problem gives you an opportunity to see what students know about tipping in restaurants. Discuss what tip they gave and why they gave that much (or that little). Be sure they are aware that the tip is added to the original meal price to determine the total amount they should pay.
2. The difficulty of this problem depends upon the tip students chose to leave in problem 1. If the tip is not an "easy fraction" of $4, Encourage students to estimate the percent of the tip.
3. **Informal Assessment** This problem assesses students' ability to estimate percents.
5. This activity is designed to avoid the confusion caused sometimes when computing with decimals. All numbers are converted to whole numbers, using cents as the unit.
6. This question stresses that percent means "so many out of 100," and that percents can be used to standardize different quantities so they can be compared directly. If students understand these two concepts, then all the goals of the unit have been reached. Be aware that some students can have an understanding of a concept without being able to verbalize their understanding in a formal way.

Section D A Final Tip 65

Each section within the unit begins with a two-page spread that describes the work students do, the goals of the section, new vocabulary, and materials needed, as well as providing information about the mathematics in the section and ideas for pacing, planning instruction, homework, and assessment.

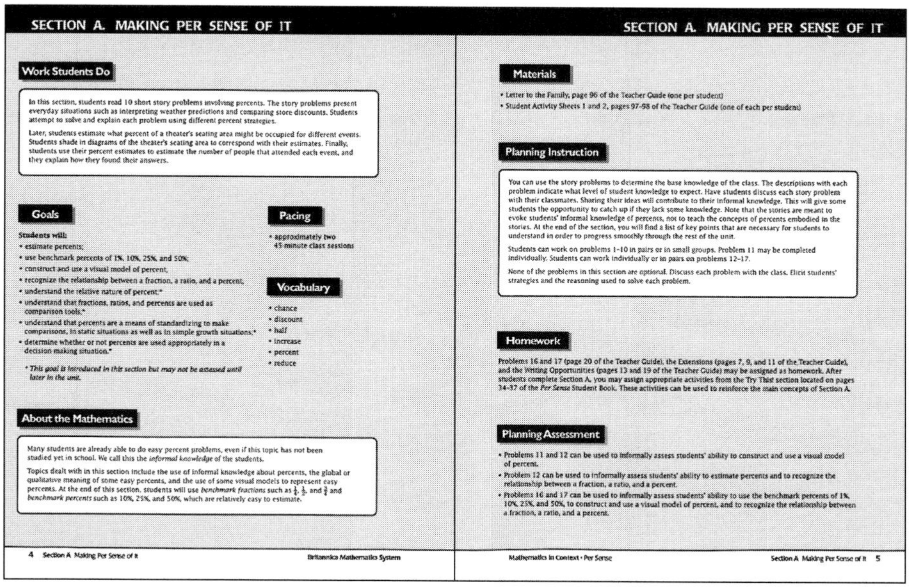
SECTION A. MAKING PER SENSE OF IT

Work Students Do

In this section, students read 10 short story problems involving percents. The story problems present everyday situations such as interpreting weather predictions and comparing store discounts. Students attempt to solve and explain each problem using different percent strategies.

Later, students estimate what percent of a theater's seating area might be occupied for different events. Students shade in diagrams of the theater's seating area to correspond with their estimates. Finally, students use their percent estimates to estimate the number of people that attended each event, and they explain how they found their answers.

Goals

Students will:
- estimate percents;
- use benchmark percents of 1%, 10%, 25%, and 50%;
- construct and use a visual model of percent,
- recognize the relationship between a fraction, a ratio, and a percent,
- understand the relative nature of percent;*
- understand that fractions, ratios, and percents are used as comparison tools;*
- understand that percents are a means of standardizing to make comparisons, in static situations as well as in simple growth situations;*
- determine whether or not percents are used appropriately in a decision making situation.*

** This goal is introduced in this section but may not be assessed until later in the unit.*

Pacing

- approximately two 45-minute class sessions

Vocabulary

- chance
- discount
- half
- increase
- percent
- reduce

About the Mathematics

Many students are already able to do easy percent problems, even if this topic has not been studied yet in school. We call this the *informal knowledge* of the students.

Topics dealt with in this section include the use of informal knowledge about percents, the global or qualitative meaning of some easy percents, and the use of some visual models to represent easy percents. At the end of this section, students will use *benchmark fractions* such as $\frac{1}{4}$, $\frac{1}{2}$, and $\frac{3}{4}$ and *benchmark percents* such as 10%, 25%, and 50%, which are relatively easy to estimate.

4 Section A Making Per Sense of It — Britannica Mathematics System

SECTION A. MAKING PER SENSE OF IT

Materials

- Letter to the Family, page 96 of the Teacher Guide (one per student)
- Student Activity Sheets 1 and 2, pages 97–98 of the Teacher Guide (one of each per student)

Planning Instruction

You can use the story problems to determine the base knowledge of the class. The descriptions with each problem indicate what level of student knowledge to expect. Have students discuss each story problem with their classmates. Sharing their ideas will contribute to their informal knowledge. This will give some students the opportunity to catch up if they lack some knowledge. Note that the stories are meant to evoke students' informal knowledge of percents, not to teach the concepts of percents embodied in the stories. At the end of the section, you will find a list of key points that are necessary for students to understand in order to progress smoothly through the rest of the unit.

Students can work on problems 1–10 in pairs or in small groups. Problem 11 may be completed individually. Students can work individually or in pairs on problems 12–17.

None of the problems in this section are optional. Discuss each problem with the class. Elicit students' strategies and the reasoning used to solve each problem.

Homework

Problems 16 and 17 (page 20 of the Teacher Guide), the Extensions (pages 7, 9, and 11 of the Teacher Guide), and the Writing Opportunities (pages 13 and 19 of the Teacher Guide) may be assigned as homework. After students complete Section A, you may assign appropriate activities from the Try This! section located on pages 34–37 of the *Per Sense* Student Book. These activities can be used to reinforce the main concepts of Section A.

Planning Assessment

- Problems 11 and 12 can be used to informally assess students' ability to construct and use a visual model of percent.
- Problem 12 can be used to informally assess students' ability to estimate percents and to recognize the relationship between a fraction, a ratio, and a percent.
- Problems 16 and 17 can be used to informally assess students' ability to use the benchmark percents of 1%, 10%, 25%, and 50%, to construct and use a visual model of percent, and to recognize the relationship between a fraction, a ratio, and a percent.

Mathematics in Context • Per Sense — Section A Making Per Sense of It 5

Assessment Activities and Solutions

Information about assessment can be found in several places in this Teacher Guide. General information about assessment is given in the Overview; informal assessment opportunities are identified on the teacher pages that face each student page; and the Assessment Activities section of this guide provides formal assessment opportunities.

Glossary

The Glossary defines all vocabulary words listed on the Section Opener pages. It includes mathematical terms that may be new to students, as well as words associated with the contexts introduced in the unit. (Note: The Student Book does not have a glossary. This allows students to construct their own definitions, based on their personal experiences with the unit activities.)

Blackline Masters

At the back of this Teacher Guide are blackline masters for photocopying. The blackline masters include a letter to families (to be sent home with students before beginning the unit), several student activity sheets, and assessment masters.

Try This! Solutions

Also included in the back of this Teacher Guide are the solutions to several Try This! activities—one related to each section of the unit—that can be used to reinforce the unit's main concepts. The Try This! activities are located in the back of the Student Book.

Unit Focus

This unit introduces students to the mathematical modeling of growth processes. Students explore several models: linear, quadratic, cubic, and exponential. They investigate the formulas and graphs for each type of growth. Students use direct and recursive formulas to gain a thorough understanding of each kind of growth. They learn how to find a model that best fits the data. The major context of this unit is the growth of plants. Students find models to fit data from real plants and use simulation and animation methods to simplify growth patterns.

Mathematical Content

- investigating linear, quadratic, cubic, and exponential functions
- using exponential notation
- using growth factors and starting values
- using recursive and direct formulas
- describing growth processes with graphs, tables, and formulas
- modeling growth processes
- using piecewise linear graphs

Prior Knowledge

This unit assumes students can do the following:

- add, subtract, multiply, and divide rational numbers, including squares;
- understand the order of operations and how to use parentheses (as in *Operations* and *Building Formulas*);
- generalize or describe a pattern using words and/or symbols and simple formulas (as in *Patterns and Symbols*, *Expressions and Formulas*, *Building Formulas*, and *Ups and Downs*);
- understand and use simple direct and recursive formulas (as in *Building Formulas* and *Ups and Downs*);
- understand linear equations and graphs that represent constant growth (as in *Ups and Downs* and *Graphing Equations*);
- read, interpret, construct, and use graphs and tables to represent a relationship and recognize the relationships among different representations: tables, graphs, and formulas (as in previous *Mathematics in Context* algebra units);
- work with exponential notation (as in *Powers of Ten*);
- use rules to compute area and volume (as introduced in *Reallotment* and *Cereal Numbers*).

This unit should be taught after the algebra strand units for the earlier grades, especially after the grade 7/8 units *Building Formulas*, *Powers of Ten*, and *Ups and Downs*, and after the grade 8/9 unit *Graphing Equations*.

Planning and Preparation

Pacing: 14–17 days

Section	Work Students Do	Pacing*	Materials
A. Plant Growth	■ investigating the growth of a number of plants using graphs and tables ■ summarizing information using a median graph	2–3 days	■ Letter to the Family (one per student) ■ Student Activity Sheet 1 (one per student) ■ See page 5 of the Teacher Guide for a complete list of materials and quantities needed.
B. Animation and Simulation	■ simulating the growth of plants using several models ■ investigating constant (linear) growth with graphs and recursive and direct formulas ■ exploring changing growth using graphs, recursive formulas, and the term *growth factor*	3 days	■ Student Activity Sheet 2 (one per student) ■ graph paper (several sheets per student) ■ number cubes (one per group)
C. Equations for Growth	■ investigating quadratic (area) and cubic (volume) growth processes ■ using graphs, tables, and formulas to study these two types of growth ■ exploring how to change the shapes of graphs and what effect this has on the formulas, and vice versa	4–5 days	■ Student Activity Sheets 3–7, (one of each per student) ■ graphing calculators or other graphing utility (one per student) ■ graph paper (several sheets per student) ■ centimeter rulers with millimeter scale (one per student)
D. Explosive Growth	■ using a device to investigate explosive (exponential) growth ■ describing explosive growth with recursive formulas and arrow language.	2–3 days	■ Student Activity Sheets 8 and 9 (one of each per student) ■ See page 63 of the Teacher Guide for a complete list of materials and quantities needed.
E. Exponential Growth	■ using exponential notation to describe exponential growth with direct formulas ■ using graphing calculators to explore exponential growth	3 days	■ Student Activity Sheet 10 (one per student) ■ See page 81 of the Teacher Guide for a complete list of materials and quantities needed.

* One day is approximately equivalent to one 45-minute class session.

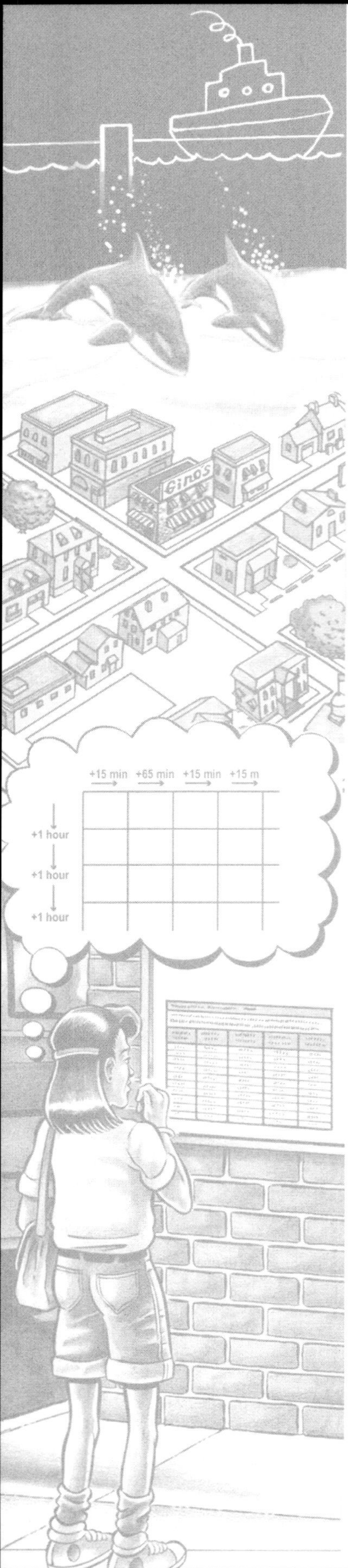

Preparation

In the *Teacher Resource and Implementation Guide* is an extensive description of the philosophy underlying both the content and the pedagogy of the *Mathematics in Context* curriculum. Suggestions for preparation are also given in the Hints and Comments columns of this Teacher Guide. You may want to consider the following:

- Work through the unit before teaching it. If possible, take on the role of the student and discuss your strategies with other teachers.
- Use the overhead projector for student demonstrations, particularly with overhead transparencies of the student activity sheets and any manipulatives used in the unit.
- Invite students to use drawings and examples to illustrate and clarify their answers.
- Allow students to work at different levels of sophistication. Some students may need concrete materials, while others can work at a more abstract level.
- Provide opportunities and support for students to share their strategies, which often differ. This allows students to take part in class discussions and introduces them to alternative ways to think about the mathematics in the unit.
- In some cases, it may be necessary to read the problems to students or to pair students to facilitate their understanding of the printed materials.
- A list of the materials needed for this unit is in the chart on page xiii.
- Try to follow the recommended pacing chart on page xiii. You can easily spend more time on this unit than the number of class periods indicated. Bear in mind, however, that many of the topics introduced in this unit will be revisited and covered more thoroughly in other *Mathematics in Context* units.

Resources

For Teachers

Books and Magazines *Mathematics Assessment: Myths, Models, Good Questions, and Practical Suggestions,* edited by Jean Kerr Stenmark (Reston, Virginia: The National Council of Teachers of Mathematics, Inc., 1991)

Assessment

Planning Assessment

In keeping with the NCTM Assessment Standards, valid assessment should be based on evidence drawn from several sources. (See the full discussion of assessment philosophies in the *Teacher Resource and Implementation Guide.*) An assessment plan for this unit may draw from the following sources:

- Observations—look, listen, and record observable behavior.
- Interactive Responses—in a teacher-facilitated situation, note how students respond, clarify, revise, and extend their thinking.
- Products—look for the quality of thought evident in student projects, test answers, worksheet solutions, or writings.

These categories are not meant to be mutually exclusive. In fact, observation is a key part of assessing interactive responses and also key to understanding the end results of projects and writings.

Ongoing Assessment Opportunities

- **Problems within Sections**
  To evaluate ongoing progress, *Mathematics in Context* identifies informal assessment opportunities and the goals that these particular problems assess throughout the Teacher Guide. There are also indications as to what you might expect from your students.

- **Section Summary Questions**
  The Summary questions at the end of each section are vehicles for informal assessment (see Teacher Guide pages 20, 34, 60, 78, and 96).

End-of-Unit Assessment Opportunities

In the back of this Teacher Guide, there are three assessment activities that, when combined, will take approximately three 45-minute class sessions. For a more detailed description of these assessment activities, see the Assessment Overview (Teacher Guide pages 98 and 99).

You may also wish to design your own culminating project or let students create one that will tell you what they consider is important in the unit. For more assessment ideas, refer to the chart on pages xvi and xvii.

Goals and Assessment

In the *Mathematics in Context* curriculum, unit goals, categorized according to cognitive procedures, relate to the strand goals and the NCTM Curriculum and Evaluation Standards. Additional information about these goals is found in the *Teacher Resource and Implementation Guide.* The *Mathematics in Context* curriculum is designed to help students develop their abilities so that they can perform with understanding in each of the categories listed below. It is important to note that the attainment of goals in one category is not a prerequisite to attaining those in another category. In fact, students should progress simultaneously toward several goals in different categories.

	Goal	Ongoing Assessment Opportunities	End-of-Unit Assessment Opportunities
Conceptual and Procedural Knowledge	**1.** recognize and describe growth patterns presented in pictures, tables, and graphs	**Section A** p. 8, #2, #4d p. 20, #17 **Section B** p. 30, #9 p. 34, #12 **Section D** p. 78, #15	Money, Money, Money, pp. 128–129 Strange Creatures, pp. 130–132 A Growth Experiment, pp. 133–134
	2. interpret and create graphs and tables representing linear, quadratic, cubic, and exponential growth processes	**Section A** p. 20, #17 **Section B** p. 32, #11 p. 34, #12 **Section C** p. 56, #18 p. 60, #20 **Section D** p. 72, #9	Money, Money, Money, pp. 128–129 Strange Creatures, pp. 130–132 A Growth Experiment, pp. 133–134
	3. interpret and create direct and recursive formulas for linear and exponential growth	**Section B** p. 30, #9 p. 32, #11 **Section D** p. 78, #15, #16 **Section E** p. 90, #8 p. 96, #15, #16	Money, Money, Money, pp. 128–129 Strange Creatures, pp. 130–132 A Growth Experiment, pp. 133–134
	4. interpret and create formulas for quadratic and cubic growth	**Section C** p. 60, #20	

	Goal	Ongoing Assessment Opportunities	End-of-Unit Assessment Opportunities
Reasoning, Communicating, Thinking, and Making Connections	**5.** understand the differences between recursive and direct formulas	**Section B** p. 32, #11 **Section E** p. 96, #15	
	6. identify common traits within each type of formula: linear, quadratic, cubic, and exponential	**Section B** p. 34, #12 **Section D** p. 78, #15 **Section E** p. 96, #15, #16	Strange Creatures, pp. 130–132 A Growth Experiment, pp. 133–134
	7. understand the similarities and differences between linear, quadratic, cubic, and exponential growth	**Section B** p. 32, #11 **Section C** p. 60, #21 **Section D** p. 78, #15	Strange Creatures, pp. 130–132 A Growth Experiment, pp. 133–134
	8. understand how changing a formula affects the graph and vice versa	**Section C** p. 48, #12 p. 60, #20	

	Goal	Ongoing Assessment Opportunities	End-of-Unit Assessment Opportunities
Modeling, Nonroutine Problem-Solving, Critically Analyzing, and Generalizing	**9.** model a growth situation by choosing an appropriate formula and justify the choice	**Section B** p. 32, #11 **Section C** p. 60, #20	Money, Money, Money, pp. 128–129 Strange Creatures, pp. 130–132 A Growth Experiment, pp. 133–134
	10. recognize that a mathematical model usually does not fit a situation perfectly	**Section A** p. 18, #15c, #15d p. 20, #17 **Section E** p. 96, #16	A Growth Experiment, pp. 133–134
	11. recognize that growth processes can be simulated with a simplified mathematical model	**Section E** p. 96, #16	Money, Money, Money, pp. 128–129 Strange Creatures, pp. 130–132 A Growth Experiment, pp. 133–134

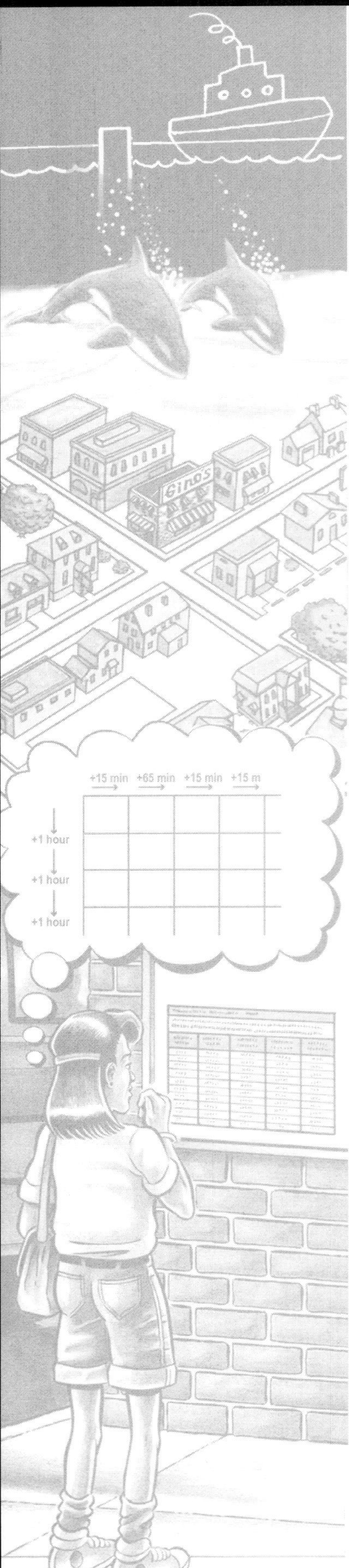

More about Assessment

Scoring and Analyzing Assessment Responses

Students may respond to assessment questions with various levels of mathematical sophistication and elaboration. Each student's response should be considered for the mathematics that it shows, and not judged on whether or not it includes an expected response. Responses to some of the assessment questions may be viewed as either correct or incorrect, but many answers will need flexible judgment by the teacher. Descriptive judgments related to specific goals and partial credit often provide more helpful feedback than percentage scores.

Openly communicate your expectations to all students, and report achievement and progress for each student relative to those expectations. When scoring students' responses try to think about how they are progressing toward the goals of the unit and the strand.

Student Portfolios

Generally, a portfolio is a collection of student-selected pieces that is representative of a student's work. A portfolio may include evaluative comments by you or by the student. See the *Teacher Resource and Implementation Guide* for more ideas on portfolio focus and use.

A comprehensive discussion about the contents, management, and evaluation of portfolios can be found in *Mathematics Assessment: Myths, Models, Good Questions, and Practical Suggestions,* pp. 35–48.

Student Self-Evaluation

Self-evaluation encourages students to reflect on their progress in learning mathematical concepts, their developing abilities to use mathematics, and their dispositions toward mathematics. The following examples illustrate ways to incorporate student self-evaluations as one component of your assessment plan.

- Ask students to comment, in writing, on each piece they have chosen for their portfolios and on the progress they see in the pieces overall.
- Give a writing assignment entitled "What I Know Now about [a math concept] and What I Think about It." This will give you information about each student's disposition toward mathematics as well as his or her knowledge.
- Interview individuals or small groups to elicit what they have learned, what they think is important, and why.

Suggestions for self-inventories can be found in *Mathematics Assessment: Myths, Models, Good Questions, and Practical Suggestions,* pp. 55–58.

Summary Discussion

Discuss specific lessons and activities in the unit—what the student learned from them and what the activities have in common. This can be done in whole-class discussions, small groups, or in personal interviews.

Connections across the *Mathematics in Context* Curriculum

Growth is one of the last grade 8/9 units in the algebra strand. The map below shows the complete *Mathematics in Context* curriculum for grade 8/9. It indicates where the unit fits in the algebra strand and where it fits in the overall picture.

A detailed description of the units, the strands, and the connections in the *Mathematics in Context* curriculum can be found in the *Teacher Resource and Implementation Guide.*

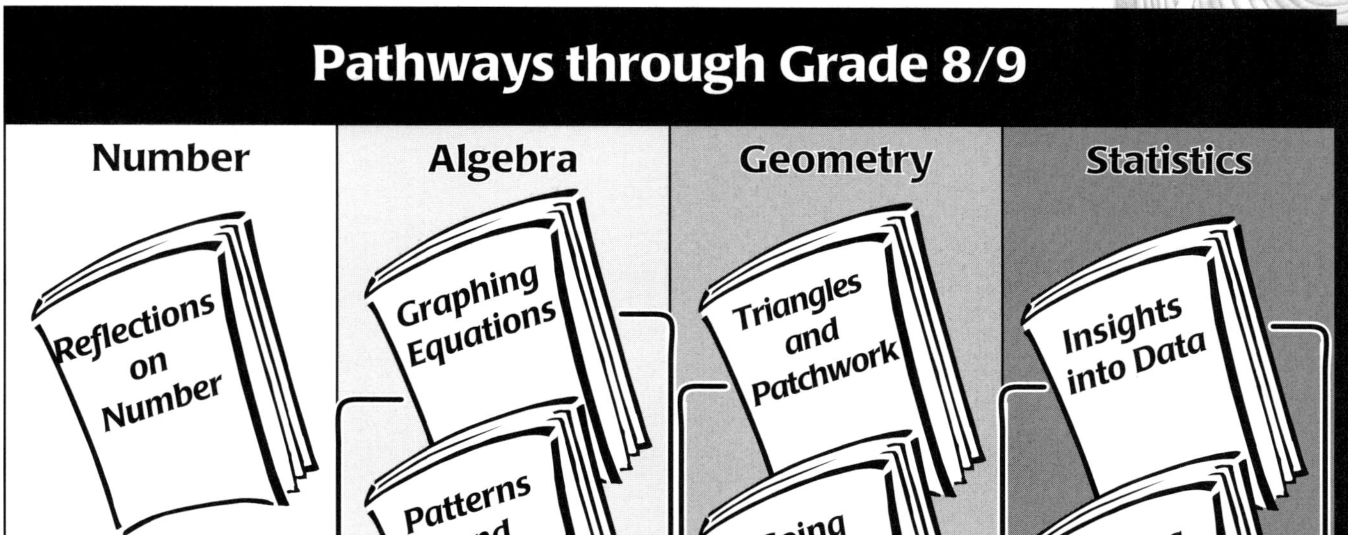

Pathways through the Algebra Strand
(Arrows indicate prerequisite units.)

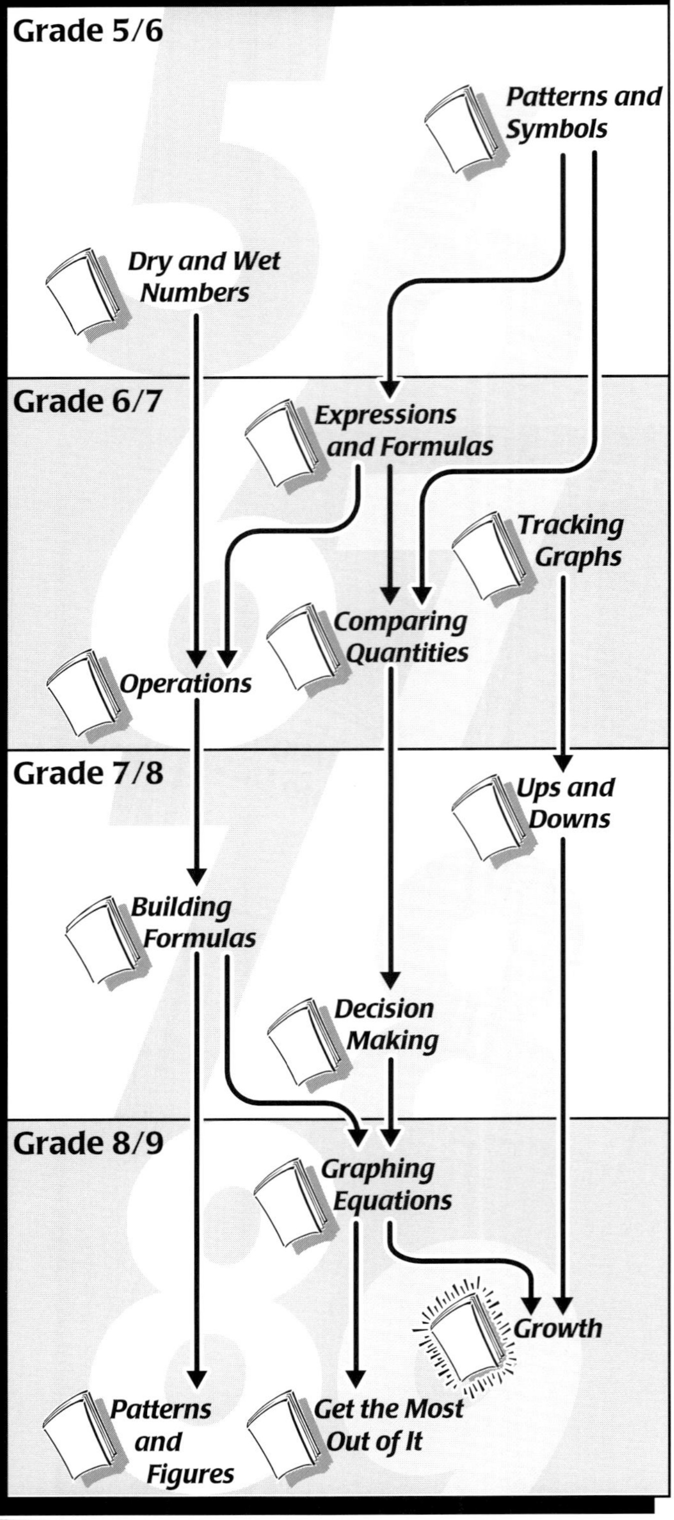

Connections within the Algebra Strand

On the left is a map of the algebra strand; this unit, *Growth,* is highlighted.

Growth can be used as the last unit in the algebra strand for grade 8/9. In this unit, many of the informal and pre-formal notions and concepts related to different types of growth processes from previous algebra units are formalized.

This algebra strand unit is preceded by the units *Building Formulas* (in which students worked with recursive and direct formulas) and *Tracking Graphs, Ups and Downs,* and *Graphing Equations* (in which the emphasis is on graphing). In these units, the linear relationship (constant growth) and its characteristics are investigated and formalized in a variety of contexts. In *Growth,* the quadratic, cubic, and exponential mathematical models are further formalized.

The *Mathematics in Context* algebra strand consists of three substrands: expressions and formulas (the mathematical notation of relationships), equations and inequalities (solving them and using them to describe situations), and graphs (as representations of processes). *Growth* is the closing unit in this third substrand, and it uses the mathematical notation and expressions developed in the first substrand.

The grade 8/9 unit *Graphing Equations* is a point in the algebra strand where all three substrands come together. The units that follow *Graphing Equations* take the equation and graph of a linear relationship as a basis to further formalize the algebra that has been studied in grades 5/6–7/8.

The more formal use of formulas and expressions (including quadratic and hyperbolic relationships) is further investigated in the unit *Patterns and Figures.*

The mathematics of solving systems of equations and linear programming is continued and further formalized in the unit *Get the Most Out of It.*

The Algebra Strand

Grade 5/6

Side Seeing
Exploring the relationship between three-dimensional shapes and drawings of them, seeing from different points of view, and building structures from drawings.

Figuring All the Angles
Estimating and measuring angles and investigating direction, vectors, and rectangular and polar coordinates.

Grade 7/8

Packages and Polygons
Recognizing geometric shapes in real objects and representations, constructing models, and investigating properties of regular and semi-regular polyhedra.

Looking at an Angle
Recognizing vision lines in two and three dimensions; identifying and drawing shadows and blind spots; identifying the isomorphism of vision lines, light rays, flight paths, and so forth; understanding the relationship between angles and the tangent ratio; computing with the tangent ratio.

Ways to Go
Reading and interpreting different kinds of maps, comparing different types of distances, progressing from one two-dimensional model to another (from a diagram to a map to a photograph to a graph), drawing graphs and networks. (*Ways to Go* is also in the statistics strand.)

Triangles and Beyond
Exploring the interrelationships of the sides and angles of triangles as well as the properties of parallel lines and quadrilaterals; constructing triangles; and using transformations to become familiar with the concepts of congruence and similarity.

Grade 6/7

Reallotment
Measuring regular and irregular areas; discovering links between area, perimeter, surface area, and volume; and using English and metric units.

Made to Measure
Measuring length (including circumference), volume, and surface area using metric units.

Grade 8/9

Triangles and Patchwork
Understanding similarity and using it to find unknown measurements for similar triangles and developing the concept of ratio through tessellation.

Going the Distance
Using the Pythagorean theorem to investigate distances, scales, and vectors and using slope, tangent, area, square root, and contour lines.

Digging Numbers
Using the properties of height, diameter, and radius to determine whether or not various irregular shapes are similar; predicting length using graphs and formulas; exploring the relationship between three-dimensional shapes and drawings of them; and using length-to-width ratios to classify various objects. (*Digging Numbers* is also in the statistics strand.)

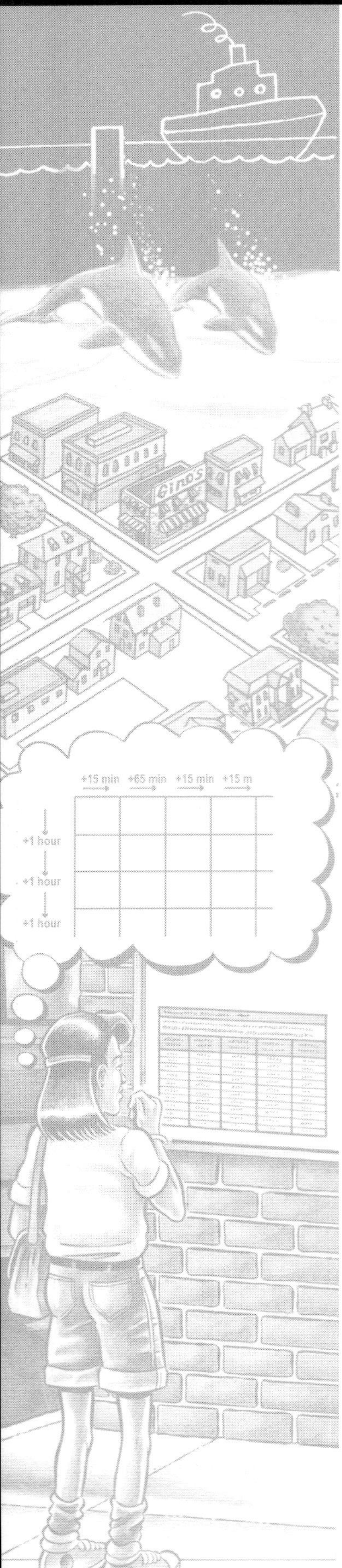

Connections with Other *Mathematics in Context* Units

Graphing is one of the three substrands in the algebra strand that focus on process, change, and growth. *Tracking Graphs* is the introductory unit within this substrand at the grade 6/7 level in which students study line graphs. In *Ups and Downs,* students use graphs to explore nonlinear, increasing and decreasing periodic functions. In *Graphing Equations,* linear relationships are formalized as students use graphing calculators to explore the equation of a line in slope and y-intercept form. In *Growth,* students investigate the differences between linear, quadratic, cubic, and exponential growth and explore the relationships between graphs and their defining formulas.

This unit builds on students' knowledge of area and volume (developed in the geometry strand) and of exponential notation and real numbers (developed in the number strand). The idea of collecting data and finding an appropriate mathematical model to describe the data is investigated in the statistics strand units.

The following mathematical topics that are included in the unit *Growth* are introduced or further developed in other *Mathematics in Context* units.

Prerequisite Topics

Topic	**Unit**	**Grade**
growing patterns	*Patterns and Symbols*	5/6
	Building Formulas	7/8
graphing processes	*Tracking Graphs*	6/7
	Ups and Downs	7/8
recursive and direct formulas	*Ups and Downs*	7/8
	Building Formulas	7/8
linear relationships	*Building Formulas*	7/8
	Ups and Downs	7/8
	Graphing Equations	8/9
area and volume	*Reallotment**	6/7
	*Cereal Numbers***	7/8
exponential notation	*Powers of Ten***	7/8
growth factors	*More or Less***	6/7

Topics Revisited in Other Units

Topic	**Unit**	**Grade**
growing patterns	*Patterns and Figures*	8/9
recursive formulas	*Patterns and Figures*	8/9
experimental growth	*Insights into Data****	8/9

* This unit in the geometry strand also helps students make connections to ideas about algebra.
** These units in the number strand also help students make connections to ideas about algebra.
*** This unit in the statistics strand also helps students make connections to ideas about algebra.

Student Materials and Teaching Notes

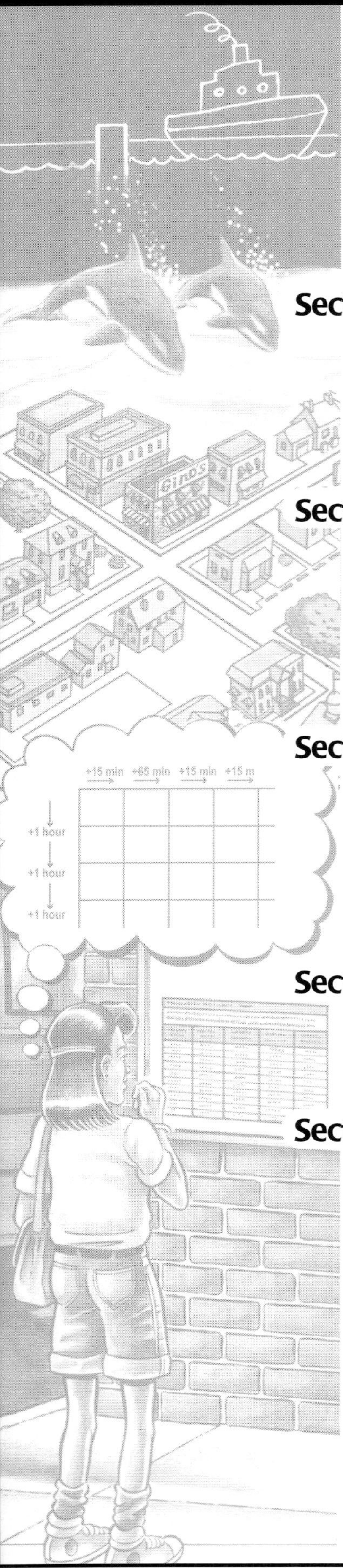

Student Book
Table of Contents

Dear Student,

Welcome to the unit *Growth.*

In this unit, you will study different types of formulas and their graphs in order to determine the best formula for describing the growth of a plant.

Have you ever planted a garden? If so, you have probably noticed that not all plants grow at the same rate. One tomato plant may shoot up right away, but by the end of the next week another tomato plant may overtake it. In this unit, you will study data on several Wisconsin fast plants and then find a way to summarize the data and describe the growth of the typical Wisconsin fast plant.

You will also create models that simulate plant growth, the area of a leaf, and the weight of a snail and find formulas that describe your models. Some of the formulas that you will work with are quadratic equations, such as $y = \frac{1}{2}x^2$, and some are cubic equations, such as $y = 0.07x^3$. You will experiment with different kinds of quadratic and cubic equations on your graphing calculator.

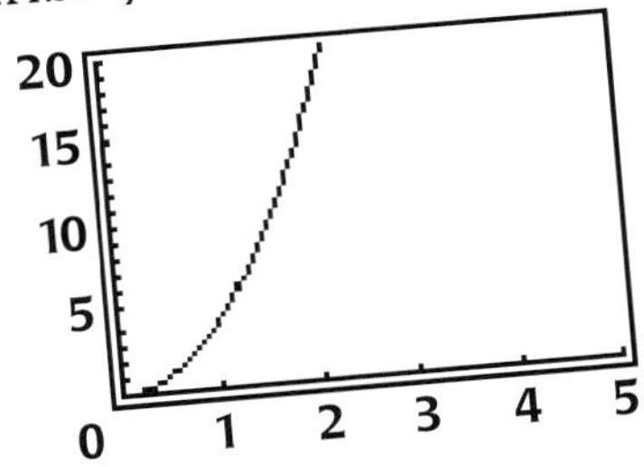

Finally, you will create a model that simulates explosive growth (growth that starts out slowly and then rapidly increases). You will discover that explosive growth is described by an exponential formula, such as $y = 20(0.4)^x$, where 20 is the starting value and 0.4 is the growth factor.

We hope you enjoy modeling these different kinds of growth.

Sincerely,

The Mathematics in Context Development Team

SECTION A. PLANT GROWTH

Work Students Do

Students study the growth of plants. They describe the explosive growth of a plant as illustrated on five occasions. They then investigate the growth of six plants over 17 days, plotting given data to compare the plants' growth and using a table to rank the plants in order from tallest to shortest. Students compare the growth of plants grown in different soils. They also explore ways to describe the typical growth of a plant. They review a new type of graph depicting data for 12 plants as unconnected points and compare this graph with those they made earlier in the section. They also compare piecewise linear graphs with graphs that are curved lines. They explain how to determine whether one soil is better than another for plant growth and use a graph to describe the growth of five plants over four weeks.

Goals

Students will:

- recognize and describe growth patterns presented in pictures, tables, and graphs;
- interpret and create graphs and tables for growth processes;
- recognize that a mathematical model usually does not fit a situation perfectly;
- recognize that growth processes can be simulated with a simplified mathematical model.*

* *This goal is introduced in this section and is assessed in other sections of the unit.*

Pacing

- approximately two to three 45-minute class sessions

Vocabulary

- line graph
- median
- piecewise linear graph

About the Mathematics

It is not possible to make predictions about the growth of natural things by making a single observation. By looking at a number of examples, it is possible to observe trends that may be used to make predictions. Care should be taken, however, since outliers can skew predictions, depending upon how generalizations are made.

Tables and graphs each have their advantages for recording growth processes: tables give exact numbers, while graphs give a better picture of the overall process. Growth is usually a continuous, smooth process without "jumps." In a graph, this is best represented by a curve and not by a piecewise linear graph (in which data points are connected by straight line segments).

Materials

- Letter to the Family, page 116 of the Teacher Guide (one per student)
- Student Activity Sheet 1, page 117 of the Teacher Guide (one per student)
- graph paper, pages 9, 13, 19, and 21 of the Teacher Guide (several sheets per student)
- transparency of graphs on Student Book page 7, page 19 of the Teacher Guide, optional (one per class)
- overhead projector, page 19 of the Teacher Guide, optional (one per class)

Planning Instruction

To begin this section, you may wish to introduce the study of the growth of plants by asking students who have planted gardens to describe the growth that occurred. Ask *Did all plants of the same type grow the same amount each week? What might account for differences in growth?*

Have students work straight through problems 2–6. Problems 11–15 should also be done at one time. At the end of the section, you may want to talk with your students about the characteristics of growth processes and about how to predict growth between and beyond measured values. During the section, students should start to realize that with a small amount of data, it is difficult to make conjectures about general growth processes.

Students could grow their own fast plants and collect the same data as are represented in the section. If you wish to do this, you will have to begin growing the plants at least 17 days prior to beginning the unit. Fast plant seeds are available through mail order via the Internet.

Students may work on problem 1 in small groups or as a class. Students may work on problems 2–6 and 11–15 in small groups. Problems 7–10 may be done individually or in small groups. Problems 16 and 17 should be done individually.

There are no optional problems in this section.

Homework

Problems 8, 9, and 10 (page 12 of the Teacher Guide) may be assigned as homework. The Extension (page 7 of the Teacher Guide) and the Writing Opportunity (page 19 of the Teacher Guide) may also be assigned as homework. After students complete Section A, you may assign appropriate activities from the Try This! section, located on pages 43–46 of the *Growth* Student Book. The Try This! activities reinforce the key mathematical concepts introduced in this section.

Planning Assessment

- Problems 2 and 4d can be used to informally assess students' ability to recognize and describe growth patterns presented in pictures, tables, and graphs.
- Problems 15c and 15d can be used to informally assess students' ability to recognize that a mathematical model usually does not fit a situation perfectly.
- Problem 17 can be used to informally assess students' ability to recognize and describe growth patterns presented in pictures, tables, and graphs; to interpret and create graphs and tables for growth processes; and to recognize that a mathematical model usually does not fit a situation perfectly.

A. PLANT GROWTH

Wisconsin Fast Plants

Michelle and Erica are growing Wisconsin fast plants (a type of wild mustard) in their biology class. The following picture shows plants that have been growing for 2, 4, 9, 13, and 18 days.

1. Based on the picture shown below, describe how Wisconsin fast plants grow.

Solutions and Samples
of student work

1. Descriptions will vary. Sample descriptions:
   - At first, the plant is just a stem and leaves. As the stem lengthens, the leaves spread out. Then the plant produces flowers.
   - For the first nine days, the plant does not grow so fast. During the next nine days, the plant grows much faster.

Hints and Comments

Overview Students are introduced to the context of fast-growing plants. They describe the explosive growth of a plant as illustrated on five occasions.

About the Mathematics One of the ways in which mathematics is used is to model situations. To fully understand a process, one often tries to find a (mathematical) model to describe it. In this unit, students work toward finding formulas to describe the growth processes of plants.

Planning Students may work on problem **1** in small groups or as a class.

Extension You may ask students to collect graphs from magazines or newspapers that depict growth processes. You can display these graphs to use as examples of the different types of growth that will be studied in this unit.

Comparing Six Plants

Wisconsin fast plants complete their life cycle in five weeks (one-fifth of the time of similar plants). Michelle and Erica are studying the plants' rate of growth. They collect the following data for six plants.

Plant Heights

	Height (in mm)					
	Day 5	Day 7	Day 10	Day 12	Day 14	Day 17
Plant A	4	7	20	47	85	138
Plant B	2	5	22	52	78	145
Plant C	3	7	28	49	150	165
Plant D	4	10	29	92	109	220
Plant E	4	11	25	82	111	155
Plant F	2	8	40	75	148	225

2. Describe the growth of the six plants listed in the above table.

3. **a.** On graph paper, show the growth of plant A by plotting points for the six height measurements given in the table.

b. Connect consecutive points with a line to make a *line graph.*

c. Your line graph can also be called a *piecewise linear graph.* Why do you think this name was given to it?

d. Estimate the height of plant A at 9 days. Explain how you found your answer.

4. **a.** Plot the data for plant B on the same graph that you used for plant A. Connect the points to make a piecewise linear graph.

b. Compare the growth of plant B to that of plant A.

c. Plot the remaining plants on the same graph in the same way.

d. Write a paragraph comparing the growth of the six plants.

2. Descriptions will vary. Sample description:

   Each plant grew throughout the 17-day period, although the plants grew at different rates. No one plant grew fastest during every time segment. Plant F was tallest at 17 days, but was not always the tallest. Most plants grew the fastest between day 12 and day 17.

3. a–b.

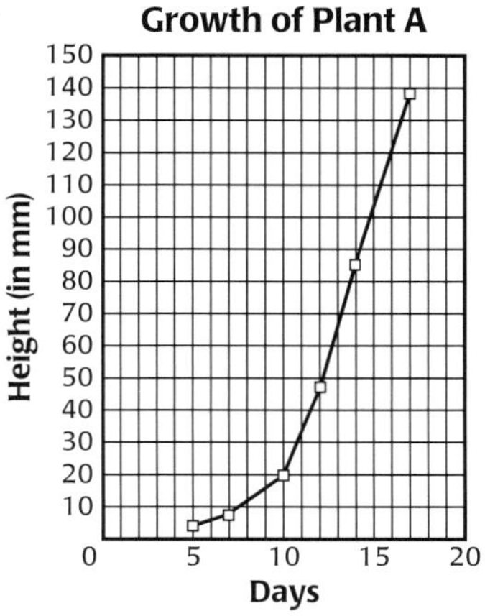

   c. Answers will vary. Sample response:

   Piecewise linear seems to be a good name because the graph is made up of line segments that represent chunks, or pieces, of time.

   d. Answers will vary. Sample estimate and explanation:

   About 16 millimeters. From day 7 until day 10, the plant grew 13 millimeters, or about 4 millimeters per day. I subtracted 4 millimeters from day 10 to find the height on day 9 (20 mm − 4 mm = 16 mm).

4. a.

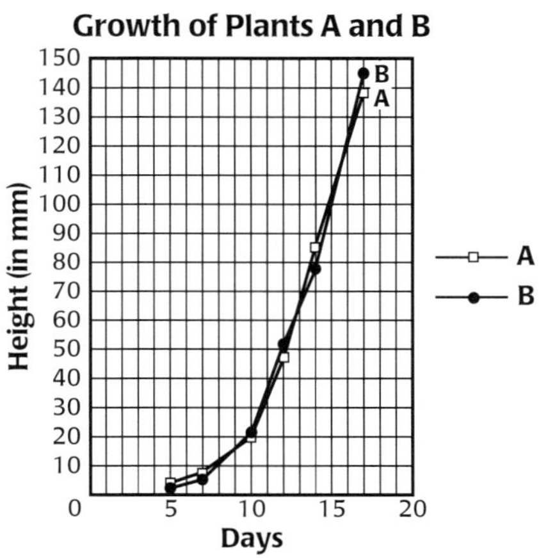

   b. Answers will vary. Sample response:

   On day 5, plant B is shorter than plant A, and it remains shorter until around day 9. From that time until around day 13, plant B is taller. From day 13 until approximately the 16th day, plant B is shorter. Then plant B is taller than plant A.

Note: See the solutions column on page 11 for answers to problems **4c–d.**

Hints and Comments

Materials graph paper (several sheets per student)

Overview Students investigate the growth of six plants over 17 days. They plot the data to compare the plants' growth.

About the Mathematics The data in the table are taken at six moments in time. A piecewise linear graph connects the data points with line segments, suggesting that growth was constant from one point to another and then suddenly became much faster or slower. Drawing a curve from one point to another represents plant growth much more realistically.

Estimating a value between two data points is called *interpolation,* and estimating a point beyond the range of the graph is called *extrapolation.*

Planning Students may work on problems **2–4** in small groups. Problems **2** and **4d** may be used for informal assessment. Discuss the different ways to depict the growth of plants.

Comments about the Problems

2. **Informal Assessment** This problem assesses students' ability to recognize and describe growth patterns presented in tables.

3. The piecewise linear graph is based on the assumption that the growth was linear and constant between measurements, as shown by the line segments connecting the points in the graph; however, overall the growth of the plants is not linear. The fact that the amount of growth varies over different time intervals is evidenced by the existing data points: they do not form a straight line when connected.

   d. Students must interpolate to determine the height at 9 days.

4. **d. Informal Assessment** This problem assesses students' ability to recognize and describe growth patterns presented in pictures, tables, and graphs.

Michelle and Erica decide to compare the growth of the six plants by ranking the plants in order from tallest to shortest for days 7, 10, 12, 14, and 17. (They choose not to rank the plants for day 5 because there was very little variability in the plants' heights on that day.)

Plant Height Rankings

Rank (from tallest to shortest)	Day 7	Day 10	Day 12	Day 14	Day 17
First	E				
Second	D				
Third	F				
Fourth	A, C				
Fifth					
Sixth	B				

5. **a.** Which plant was the tallest on day 7? How tall was it?
   b. Why is the box for the fifth-tallest plant on day 7 blank?
   c. Copy the above table into your notebook and rank the plants from tallest to shortest for days 10, 12, 14, and 17.
   d. Based on your rankings, which of the six plants grew the fastest? Explain.

6. Do you think that the rankings from day 17 will remain the same as the plants continue to grow? Explain.

4. **c.**

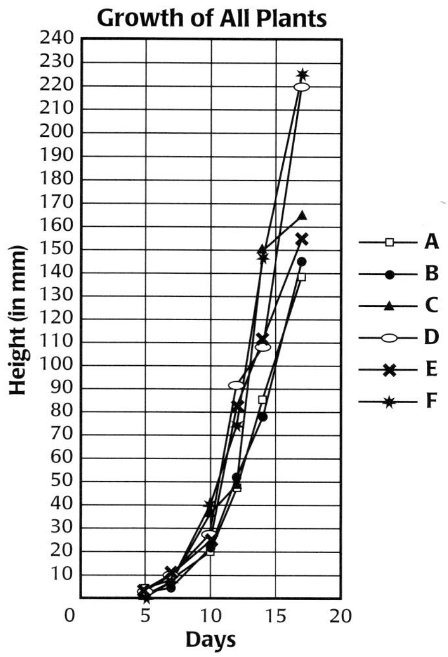

d. Answers will vary. Sample description:

The growth patterns of the plants are different. Overall, plants A and B were the slowest to grow, and plant F grew the fastest. With each day's measurement, however, the ranking of an individual plant with respect to the others changes.

5. **a.** plant E; 11 millimeters tall

b. Plants A and C are tied for fourth.

c.

Plant Height Rankings

Rank (from tallest to shortest)	Day 7	Day 10	Day 12	Day 14	Day 17
First	E	F	D	C	F
Second	D	D	E	F	D
Third	F	C	F	E	C
Fourth	A, C	E	B	D	E
Fifth		B	C	A	B
Sixth	B	A	A	B	A

d. Answers and explanations will vary. Sample response:

Based on the ranking of the final day, plant F was the fastest-growing plant. Plant F was also the only plant that was ranked first at two measurement times.

6. Answers will vary. Sample responses:

- Probably not. It appears that the older the plants are, the more random the growth pattern becomes.
- Although we know that the growth of the plants is rapid, the predictability of that growth is poor. The growth rate is not constant and is definitely not the same for each plant.

Hints and Comments

Overview Students continue the investigation of the growth of the six plants. They use a table to rank the plants in order from tallest to shortest.

Planning Students may continue working in small groups on problems **5** and **6.**

Comments about the Problems

5. b. The convention is that when there is a tie in any ranking, the item ranked below the items that are tied is moved down one place.

DIFFERENT SOILS

John grew one plant in soil that is different than the soil that was used for plants A–F. The following table shows the growth of this plant, called plant Z.

Plant Z

Day	5	7	10	12	14	17
Height (in mm)	4	8	11	40	65	120

7. Do you think this soil is less suitable for growing Wisconsin fast plants than the soil used for plants A–F? Explain.

Plant G, whose measurements are shown below, was grown in the same soil as plants A–F.

Plant G

Day	5	7	10	12	14	17
Height (in mm)	3	3	18	34	87	98

8. With this additional information, would you change your answer to problem **7?** Explain.

9. Study the table on page 2 and your graph of plants A–F. Is it possible to identify one plant as a "typical" plant? Explain.

The following table shows the growth of another Wisconsin fast plant up to 12 days. It was grown in the same soil as plants A–F.

Plant Growth

Day	5	7	10	12
Height (in mm)	3	6	25	80

10. Using the data and graphs for plants A–F, can you predict the height of this plant at 14 days? Explain.

7. Answers and explanations will vary. Sample response:

   Compared with the other plants, this plant's growth is slower. On day 17, it was 18 millimeters smaller than the smallest plant that Michelle and Erica studied, and from day 10 on, this plant did not grow as tall as the others. If all other conditions were identical, then it would seem that the soil is less suitable.

8. Answers will vary, depending on the response to problem **7.** Students should note that plant Z is no longer an outlier, so the soil in which it was grown is not necessarily less suitable.

9. Answers and explanations will vary. Sample answers:

   - With such great variability, you cannot identify any one of the plants as typical without some other evidence.
   - Plant E is somewhat in the middle of the graph and can therefore be identified as "typical."

10. Answers and explanations will vary. Some students may predict the height based on a plant that has a similar growth rate, such as plant E. At 14 days, plant E was 111 millimeters tall.

Hints and Comments

Materials graph paper, optional (several sheets per student)

Overview Students compare the growth of plants grown in different soils.

About the Mathematics The graph and table used here to show growth each have advantages and disadvantages. In a table, the actual numbers can be read off, while a graph provides a picture of the whole process over a period of time.

Planning You can introduce this page with a brief class discussion about conditions in which plants grow. When the conditions are different, plants usually grow differently. Students may work on problems **7–10** individually or in small groups. When students work on these problems, make sure they have their graphs for plants A–F at hand. Problems **8–10** may be assigned as homework.

Comments about the Problems

7. Based on data for only one plant, it is hard to say something for all plants in that different soil. Students should realize that they need to be careful with making generalizations based on one or a few individuals.

7–8. Students may graph the heights of the plants or just compare the tables.

8–10. Homework These problems may be assigned as homework.

Interdisciplinary Connection Have students talk with their science teacher to find out what factors influence the growth of plants. They can write a couple of paragraphs about the conditions in which plants grow. The following factors are important: the quality of the soil, the amount of water, the amount of sunlight, and the temperature.

Typical Growth

So far, you have seen how the growth of Wisconsin fast plants varies. It is difficult to accurately predict the growth of a single plant, but it is possible to describe the typical growth of a plant.

Use the graphs of plants A–F that you plotted in problems **3** and **4** to answer problem **11.**

11. a. Shade a band between the highest and the lowest curves on your graph that shows the range of growth for all of the plants from day 5 through day 17.

b. What can you determine about the growth of Wisconsin fast plants if your band is very wide?

A band, such as the one that you made for problem **11a,** shows the trends in the growth of Wisconsin fast plants. You can study the trends in growth to make predictions about how a typical plant will grow.

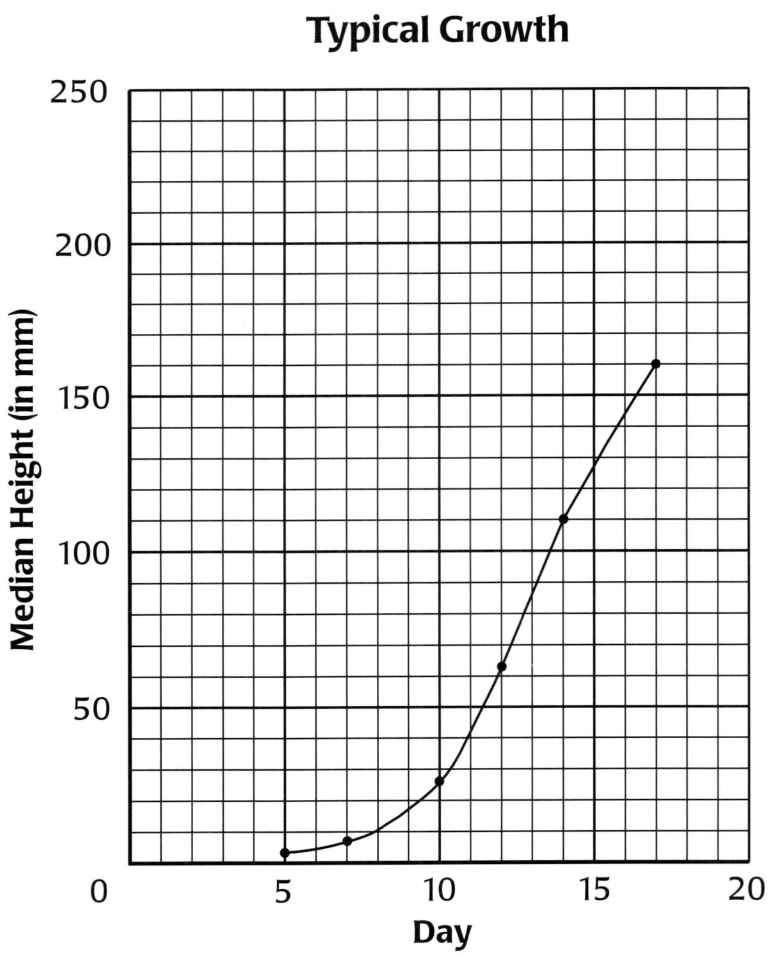

Michelle and Erica decide to summarize the information shown by the band with a curve. They find the median height for each day and then make the graph on the left to show the typical growth of a Wisconsin fast plant.

12. Is Michelle and Erica's graph showing typical growth misleading? Explain.

11. a.

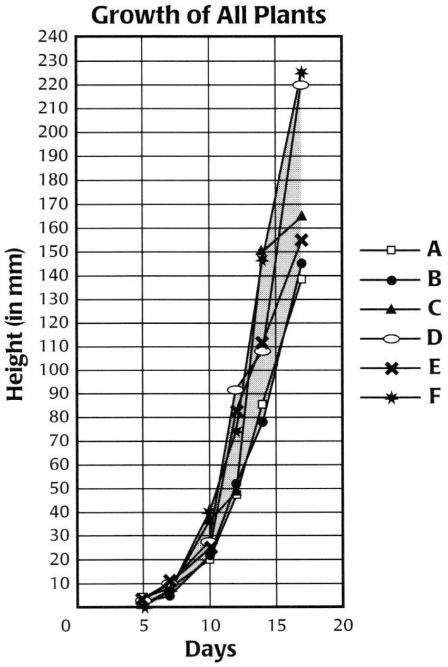

b. that there is a wide variation in rates of growth

12. Answers will vary. Sample response:

It is misleading in that it does not represent the growth of any one plant. Michelle and Erica have lost the exact values for each measurement and the information about the *range* of growth, but they have gained a visual picture of the overall growth pattern.

Materials graphs plotted in problems **3** and **4**

Overview Students explore ways to describe the typical growth of a plant using the information they have been given for the growth of plants A–F.

Planning After introducing students to the idea of a trend, you can have them work in small groups on problems **11** and **12.**

Comments about the Problems

11. A wide band means that there was a great deal of variation in the growth of the plants represented. It is not as useful for making a generalization than a narrow band would be.

One way to narrow the band is to disregard the highest and lowest data points for each interval. Another is to average the three highest and the three lowest readings at each interval.

12. Using the median (middle value) of each range of values at each time period gives outliers less influence. This is different than using the mean, or average, value.

As students should remember from the statistics units, any time data is summarized, you lose information about the individual cases, but you get a clearer overall picture. Students need to be aware of this effect of summarizing data.

In addition to the data for plants A–F, Michelle and Erica collect data for six more plants (plants G–L). The following table shows the data for all 12 plants.

	Height (in mm)					
	Day 5	Day 7	Day 10	Day 12	Day 14	Day 17
Plant A	4	7	20	47	85	138
Plant B	2	5	22	52	78	145
Plant C	3	7	28	49	150	165
Plant D	4	10	29	92	109	220
Plant E	4	11	25	82	111	155
Plant F	2	8	40	75	148	225
Plant G	3	3	18	34	87	98
Plant H	5	16	38	59	127	205
Plant I	6	6	26	52	126	190
Plant J	5	10	37	84	128	147
Plant K	4	5	27	60	64	188
Plant L	3	9	42	49	88	127

Michelle and Erica make the following graph to display the growth data for all 12 plants.

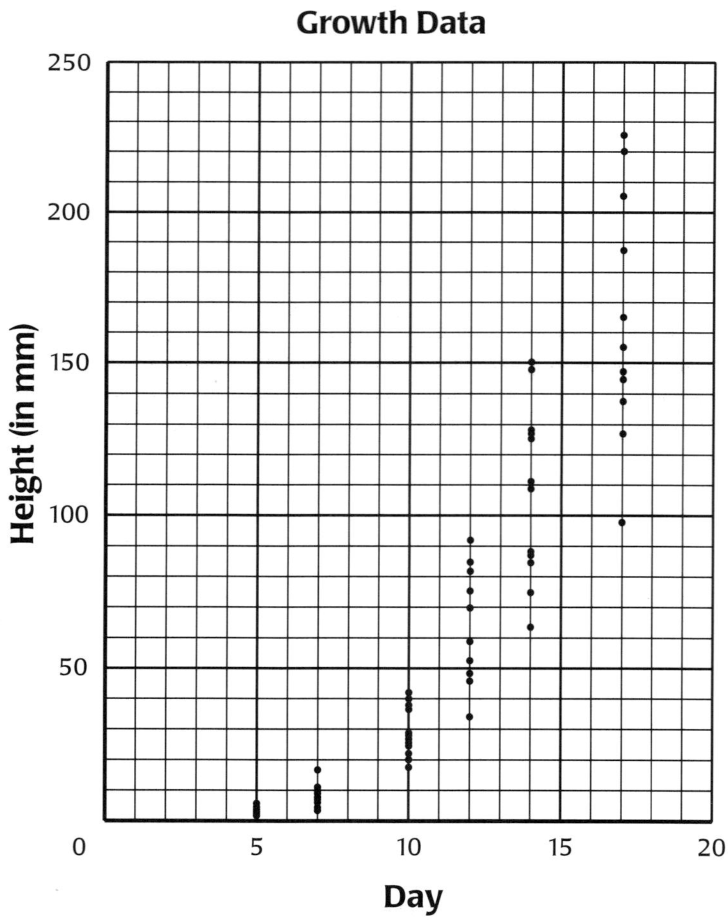

13. Compare this new type of graph with the piecewise linear graphs of plants A–F.

13. Comparisons will vary. Sample comparison:

In this new graph, the data points are not identified as belonging to particular plants, so you cannot follow the growth of individual plants. Without the lines of the piecewise linear graph, however, the range and distribution of the data are easier to see.

Overview Students review a new type of graph depicting data for 12 plants as unconnected points. They compare this graph with those they made earlier in the section.

Planning Students may continue working in small groups on problem **13.**

Comments about the Problems

13. Neither graph is a perfect model of the growth of the typical plant. The piecewise linear graph has the advantage that you can easily see the growth of each individual plant.

Extension Have students look in the table and find a plant whose growth they think is the most constant and a plant whose growth is the "strangest."

14. On Michelle and Erica's graph on **Student Activity Sheet 1,** sketch a curve that summarizes the data.

15. a. On graph paper, plot the median heights for plants A–L for each day. Connect the points to make a piecewise linear graph.

b. On another sheet of graph paper, plot the median points again. This time connect the points with a curve.

c. Which graph better represents plant growth: the piecewise linear graph or the curve? Explain.

d. Compare your graph for part **b** with other students' graphs. Are there any differences in the curves? Explain.

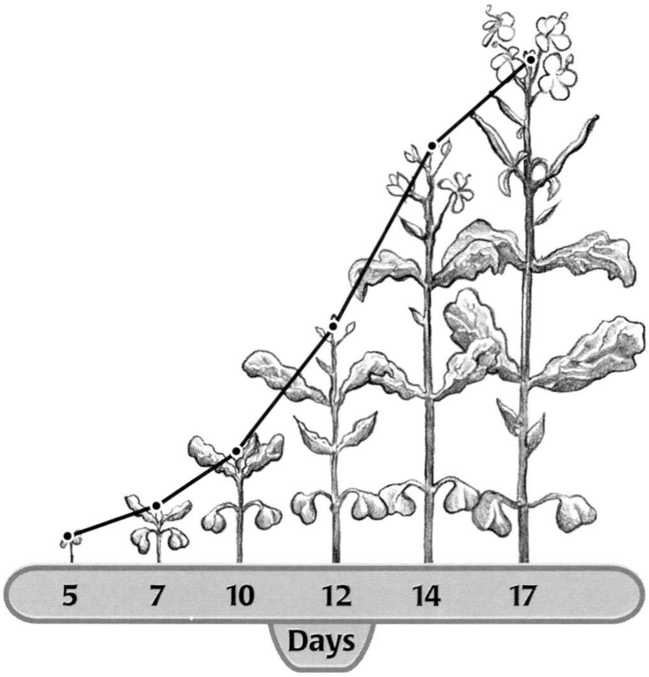

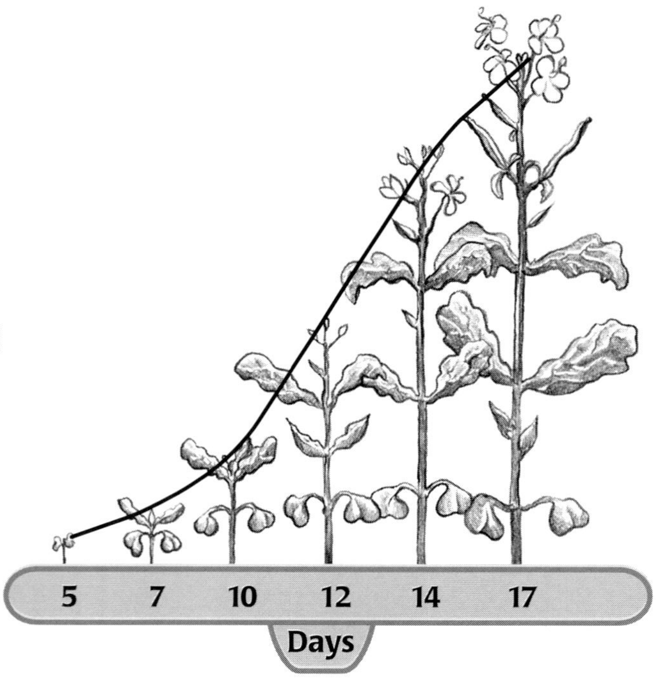

14. Graphs will vary. Sample graph:

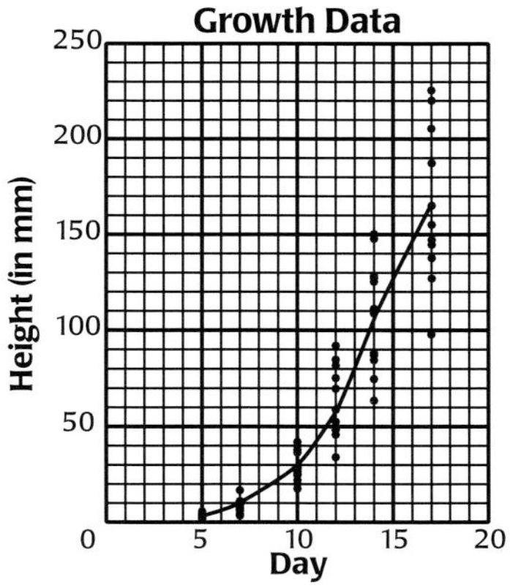

15. a.

	Day 5	Day 7	Day 10	Day 12	Day 14	Day 17
Median Height	4 mm	7.5 mm	27.5 mm	55.5 mm	110 mm	160 mm

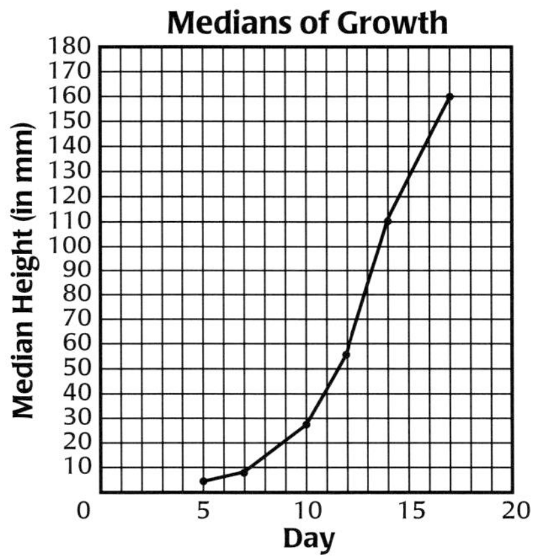

b.

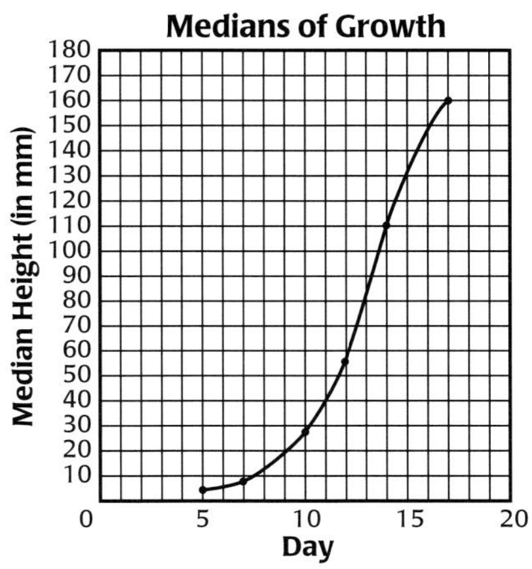

c. Answers and explanations will vary. Sample response:

The curve represents the growth better because the piecewise linear graph suggests that there are "jumps" in the growth. The curve does not show jumps.

d. Comparisons will vary, depending on the curves drawn by students for part **b.**

Materials Student Activity Sheet 1 (one per student); graph paper (several sheets per student); transparency of graphs on Student Book page 7, optional (one per class); overhead projector, optional (one per class)

Overview Students continue comparing curves with piecewise linear graphs.

Planning Students may work in small groups on problems **14** and **15.** When students have finished, discuss the differences between the graphs and how to make these graphs. The graphs in the solutions column illustrates the differences between a piecewise linear graph (**15a**) and a curve (**15b**). You may wish to put these two graphs on an overhead transparency and use this transparency in the class discussion. Problems **15c** and **15d** may be used as informal assessments.

Comments about the Problems

14. Students can do this by just looking at the graph and trying to find a curve that fits best.

15. a. Finding the median heights is finding the middle values. This works best when the table is used. When there is no middle value (that is, when there are an even number of data points), the median is the average of the two middle values.

c–d. Informal Assessment These problems assess students' ability to recognize that a mathematical model usually does not fit a situation perfectly.

Writing Opportunity You may wish to have students write their answers to problems **15c** and **15d** in their journals.

Summary

In this section, you studied data on the heights of several plants at different times. You used three methods to summarize the data:

- ranking the plants in order from tallest to shortest to compare them,
- drawing a band on a graph of the data to show the spread, and
- graphing the median for each day's data to show a measure of the centers.

One advantage of summarizing data is that you are better able to make generalizations and predictions. One disadvantage of summarizing is that you lose information about individual plants.

Summary Questions

16. In problem **7** on page 4, you considered a plant grown in a different soil. At that time, you compared the new plant to plants A–F. Explain how you would decide whether one soil is better than another.

17. The following illustration shows the growth of five plants.

a. Measure the height of each plant for each time period and record the data in a table.

b. Graph the height data from part **a.**

c. Use your answers from parts **a** and **b** to describe the growth of a typical plant of this variety.

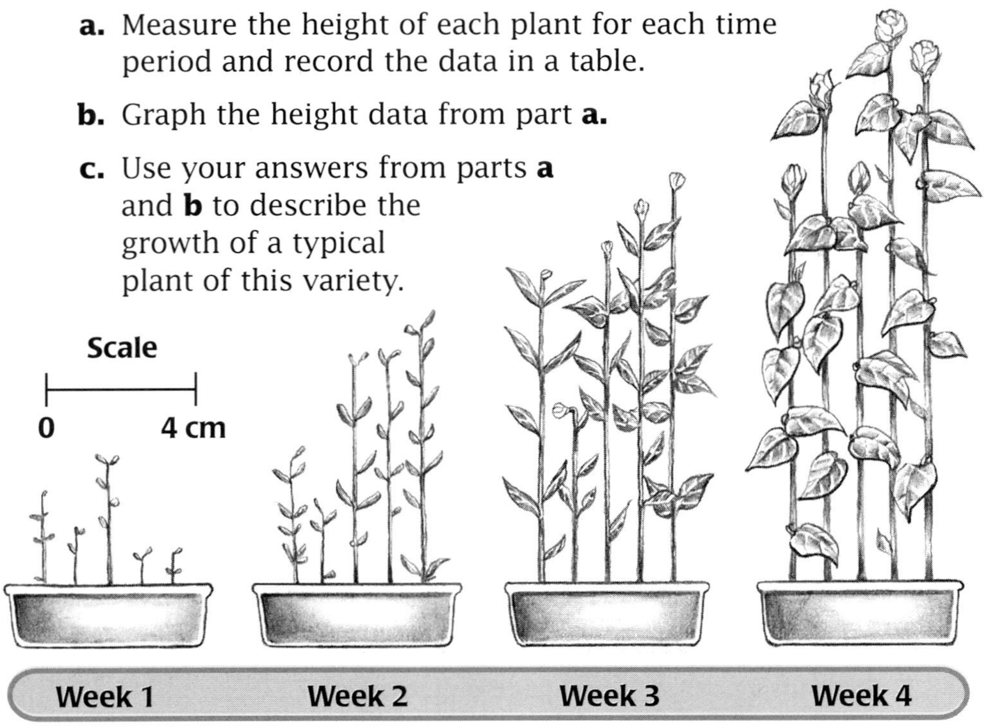

Solutions and Samples
of student work

16. Explanations will vary. Sample explanation:

You would compare the "average" growth of a number of plants in one soil to the average growth of a number of plants in another soil.

17. a. **Plant Heights**

Plant	Week 1	Week 2	Week 3	Week 4
A	2.4 cm	3.6 cm	8.2 cm	11 cm
B	1.5 cm	2.2 cm	4.6 cm	13.4 cm
C	3.4 cm	6 cm	9 cm	11 cm
D	1 cm	6.1 cm	10 cm	15 cm
E	1 cm	7.2 cm	10.6 cm	14.2 cm

b. Graphs may vary. Students may create a piecewise linear graph or a graph similar to the one on Student Book page 6. Sample graph:

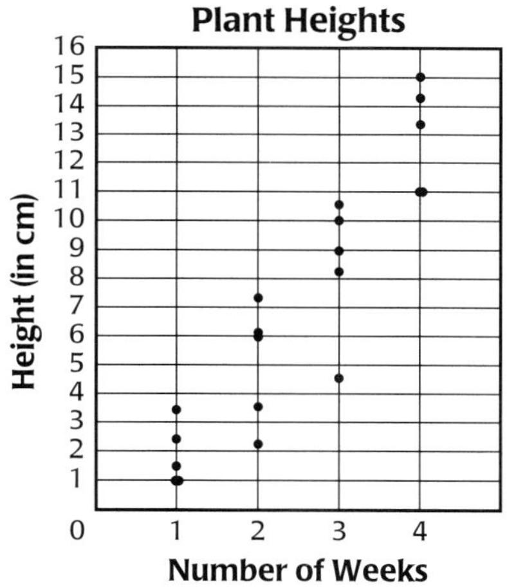

c. Answers will vary. Sample response:

The median weekly heights for the five plants studied are 1.5 cm, 6 cm, 9 cm, and 13.4 cm, which gives an average increase of around 4 cm. This is a fairly consistent growth pattern.

Hints and Comments

Materials graph paper (one sheet per student)

Overview Students read the Summary, which reviews the main concepts covered in this section. They explain how to determine whether one soil is better than another for plant growth and graph and describe the growth of five plants over four weeks.

About the Mathematics The mathematics in this section includes the following concepts and ideas: piecewise linear graphs, curved graphs, rankings, median and median graphs, the advantages and disadvantages of summarizing information, and estimating growth between measured values (interpolation) and beyond measured values (extrapolation).

Planning You can use the Summary text and the comments in the About the Mathematics section above to summarize this section. Students should work on problems **16** and **17** individually. Problem **17** can be used as an informal assessment.

After students complete Section A, you may assign appropriate activities from the Try This! section, located on pages 43–46 of the *Growth* Student Book, for homework.

Comments about the Problems

16. Students use the knowledge gained in this section to design a mathematical experiment. They should note that more than one plant would be desired and that they would have to control conditions for the experiment.

17. **Informal Assessment** This problem assesses students' ability to recognize and describe growth patterns presented in pictures, tables, and graphs; to interpret and create graphs and tables for growth processes; and to recognize that a mathematical model usually does not fit a situation perfectly.

Work Students Do

In this section, students model the growth of plants through animation. To find a mathematical model for plant growth, they start with some basic rules and compare the results to what they already know about plant growth. The first mathematical model explored is linear growth. Students first construct a model of a plant with pictures that represent the stages in the plant's growth. They then complete tables and make graphs to illustrate the given growth pattern. Students work with and formulate NEXT-CURRENT and direct formulas that summarize the rules of the given linear growth pattern. To make the plant growth model more realistic, students investigate mathematical models involving growth factors. They explore the idea of variability and model it using random numbers.

Goals

Students will:

- recognize and describe growth patterns presented in pictures, tables, and graphs;
- interpret and create graphs and tables representing linear, quadratic, cubic, and exponential growth processes;
- interpret and create direct and recursive formulas for linear and exponential growth;
- understand the differences between recursive and direct formulas;
- identify common traits within each type of formula: linear, quadratic, cubic, and exponential;
- understand the similarities and differences between linear, quadratic, cubic, and exponential growth;
- model a growth situation by choosing an appropriate formula and justify the choice;
- recognize that a mathematical model usually does not fit a situation perfectly;*
- recognize that growth processes can be simulated with a simplified mathematical model.*

* *These goals are addressed in this section and are assessed in other sections of the unit.*

Pacing

- approximately three 45-minute class sessions

Vocabulary

- animation
- changing growth
- direct formula
- growth factor
- linear or constant growth
- next-current formula

About the Mathematics

Simple mathematical models that involve recursive and direct formulas can be used to simulate plant growth. A linear (or constant) plant growth model can be described by the recursive formula NEXT = CURRENT + constant, or by the direct formula height = starting height + (growth amount × number of time periods). For example, consider the formula NEXT = CURRENT + 20. If a plant is currently 40 cm tall, it will be 60 cm tall after one week, 80 cm tall the following week, and so on. While this simple mathematical model is useful, much natural plant growth is not constant.

One way to make the mathematical model more realistic is to introduce a changing growth pattern in which the amount of growth in one period is dependent on that of the previous period. Changing growth can be described using the recursive formula NEXT = CURRENT × growth factor.

Materials

- Student Activity Sheet 2, page 118 of the Teacher Guide (one per student)
- graph paper, pages 27, 31, 33, and 35 of the Teacher Guide (several sheets per student)
- number cubes, page 33 of the Teacher Guide (one per group of students)

Planning Instruction

You might introduce this section with a brief class discussion about the growth rate of your students. Ask questions such as *Do you think that you grow at the same rate every day?* [No, adolescents do not grow at a constant rate; human growth rate varies until adulthood, at which point the growth rate stops.] *Is there ever a time in a person's life in which the growth rate is constant?* [Yes. When a person reaches adulthood, the growth rate is constant (zero).] You might also discuss the different representations of growth patterns (table, graph, and formula) and how they relate to each other. If time permits, you might also discuss the differences between constant growth and changing or variable growth and the differences between direct and NEXT-CURRENT formulas.

Students may work on problems 1–6 and 10 in small groups. Students may work individually on problems 7–9, 11, and 12.

Problem 10 is optional. If time is a concern, you may omit this problem or assign it as homework.

Homework

Problems 4 (page 26 of the Teacher Guide) and 7–9 (page 30 of the Teacher Guide) may be assigned as homework. Also, the Extension activities (pages 27, 31, and 33 of the Teacher Guide) and the Writing Opportunity (page 35 of the Teacher Guide) may be assigned as homework. After students complete Section B, you may assign appropriate activities from the Try This! section, located on pages 43–46 of the *Growth* Student Book. The Try This! activities reinforce the key mathematical concepts introduced in this section.

Planning Assessment

- Problem 9 can be used to informally assess student's ability to recognize and describe growth patterns presented in pictures and to interpret and create recursive formulas for exponential growth.
- Problem 11 can be used to informally assess students' ability to interpret and create graphs and tables representing linear and exponential growth processes and to interpret and create direct and recursive formulas for linear and exponential growth. It also assesses their ability to understand the differences between recursive and direct formulas; to understand the similarities and differences between linear and exponential growth; and to model a growth situation by choosing an appropriate formula and justify the choice.
- Problem 12 can be used to informally assess students' ability to recognize and describe growth patterns presented in pictures, tables, and graphs; to interpret and create graphs and tables representing linear growth processes; and to identify common traits within linear formulas.

B. ANIMATION AND SIMULATION

How Will the Plant Growth Continue?

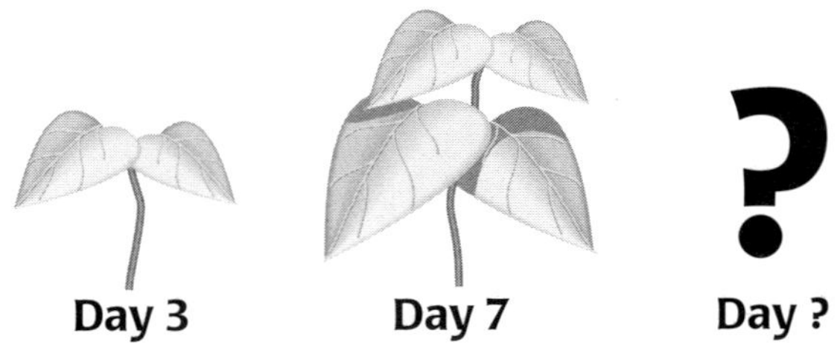

In Section A, you learned that it is difficult to predict the growth of an individual plant, but you can model the growth of a group based on information gathered from individual plants.

In this section, you will consider a new challenge. Instead of studying the growth of actual plants, you will model the growth of a plant through animation. You will perform the animation by hand. Biologists, however, do animation using computer technology.

Your first goal is to construct a model of a plant with pictures. The pictures should represent the stages in the plant's growth. Here is one model.

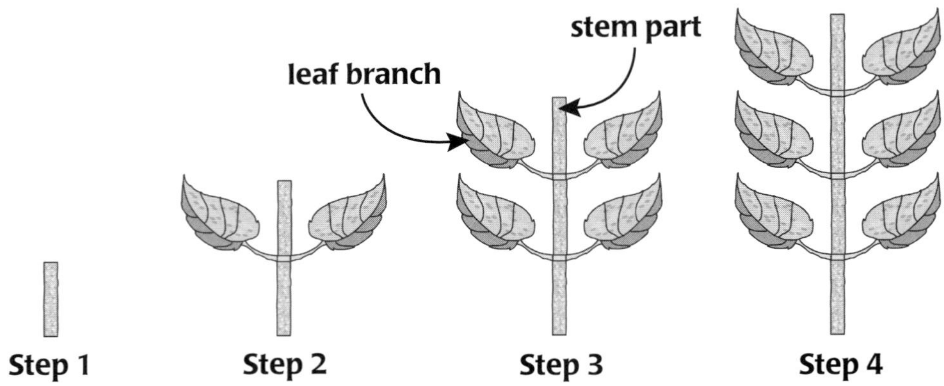

These are the building blocks used to make the model.

A flower eventually appears at the top of the stem.

1. **a.** Draw a picture of step 5 and describe the growth process for this plant. (Step 5 does not include a flower.)

b. Following the above model, draw a picture of a plant with seven stem parts and a flower. Be sure to add the appropriate number of leaf branches.

2. Suppose that for a particular plant, each stem part measures 3 centimeters, the total stem length is 60 centimeters, and there is a flower.

a. How many steps are needed to create the entire plant? Explain.

b. How many leaf branches are there? Explain.

Solutions and Samples
of student work

1. Drawings should show an additional pair of leaves and the stem extended accordingly.

a.

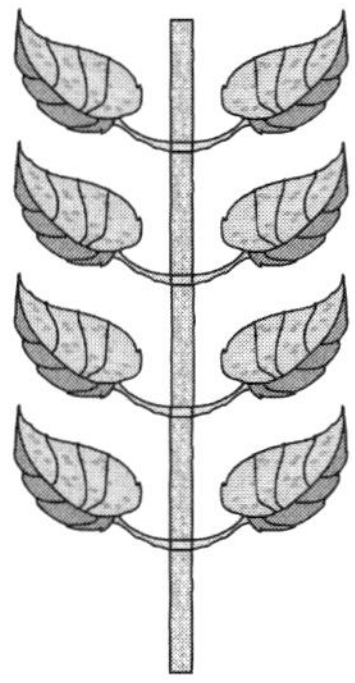

b.

2. **a.** 21 steps. Explanations will vary. Sample explanation:

   The total stem length is 60 cm, and each stem part is 3 cm. One stem part is added in each step. So, I divided to find how many stem parts add up to make a total of 60 cm: $60 \div 3 = 20$ stem parts. The flower takes one step, so 21 steps are needed.

   b. 19 leaf branches. Explanations will vary. Sample explanation:

   The total number of leaf branches is one less than the total number of stem parts, since the first stem part (in step one) has no leaf branch. So, there are $20 - 1 = 19$ leaf branches.

Hints and Comments

Overview Students begin to model plant growth through animation. They construct a plant model with pictures that follows the given rules.

About the Mathematics To gain insight in real-life growth processes, mathematical models can be useful. One often starts with a simple model that can be described by a rule or formula. The model is often later adjusted to better reflect the changing growth patterns in real-life situations by introducing growth factors. Variability can also be introduced by using chance to determine the growth during each growth period.

Planning You might discuss the meanings of the terms *animation* and *simulation* and talk about how the plant growth process that students investigated in Section A will be simplified here using a mathematical model. Students may work on problems **1** and **2** in small groups. Be sure to discuss students' answers to both problems.

Comments about the Problems

1. Students do not need to make neat drawings of the stems and leaves at each step of the plant growth. Simple sketches will suffice.

2. Students may use a variety of strategies to find the number of steps needed to create the entire plant. Some students may make a table to reflect the plant height at each step to determine when the length of the stem is 60 centimeters, although this method is rather time-consuming. Other students may construct a graph that reflects the plant height at each step. A few students may intuitively create and use a formula to describe the plant growth, although this is not expected at this point in the unit. Formulas are introduced later in this section.

   Most students will probably use some reasoning strategy. Make sure students understand that the number of steps is more than the number of stem parts when a flower is added at the top of the plant and that the number of leaf branches is one less than the number of steps.

CONSTANT GROWTH

Let's look at another plant that follows the same growth model as the one described on page 9. Each day this plant grows one more step. The length of each stem part is 12 millimeters, and the plant does not have a flower.

3. a. On graph paper, graph the height of this plant for each day over a 10-day period. Be sure to include day 0 in your graph, when the plant has a height of 0 millimeters.

b. Connect consecutive points on your graph with line segments to make a line graph and predict how tall the plant will be at the end of 10.5 days.

c. The type of growth displayed by this plant is called *linear growth.* Why do you think this name was given to it?

d. Write a direct formula for the height of the plant in terms of the number of days elapsed.

For a real plant, it can take many days before the growth pattern becomes linear. For this reason, measurements are not started when the seed first begins to germinate.

Imagine a plant that has a height of 30 millimeters when you first observe it and then grows continuously at a rate of 8 millimeters per day.

4. a. In your notebook, copy and complete the following table of the heights over a 10-day period.

Day	0	1	2	3	4	5	6	7	8	9	10
Plant Height (in mm)	30										

b. On graph paper, make a graph of the heights.

c. Write a direct formula for the height in terms of the number of days that have passed since you started observing this plant.

d. Draw a series of pictures of this plant.

A good model of plant growth should resemble the growth of a real plant.

5. Is the model of plant growth that you have been using in this section realistic? Why or why not?

Solutions and Samples
of student work

3. a–b.

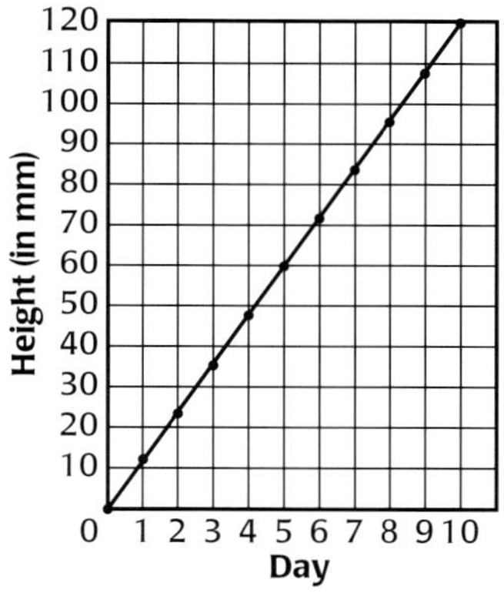

The plant will be 126 mm tall after 10.5 days

c. Explanations will vary. Sample explanation:

This kind of growth is called linear growth because the graph will always be a straight line.

d. The direct formula is $H = 12 \times D$, where H = height in millimeters and D = days elapsed.

4. a.

Day	0	1	2	3	4	5
Plant Height (in mm)	30	38	46	54	62	70

Day	6	7	8	9	10
Plant Height (in mm)	78	86	94	102	110

b.

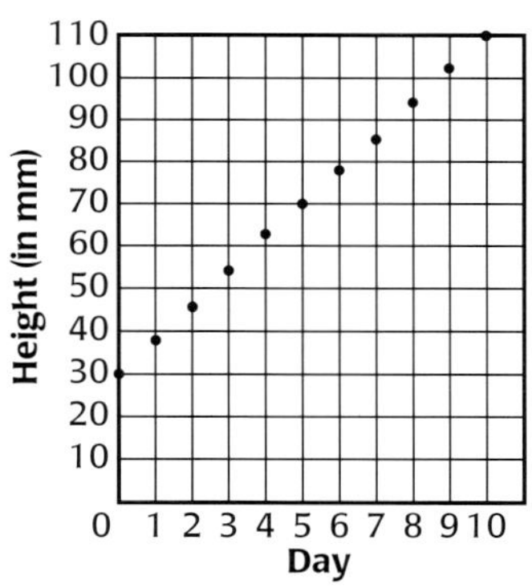

c. The direct formula is $H = 30 + 8 \times D$, where H = height in millimeters and D = days elapsed since the first observation.

d. Drawings will vary, depending on the growth stages drawn by students. Drawings of the plant growth before a plant reaches a height of 30 mm could show irregular growth. Drawings of the plant growth after a plant reaches 30 mm should show linear growth.

5. Answers and explanations will vary. Sample response:

The model of plant growth used in Section B is not realistic since most plant growth is not linear. Plant growth variation may be due to differences in soil, rainfall, climate, and so on.

Hints and Comments

Materials graph paper (two sheets per student)

Overview Students investigate two models of constant growth that are represented by a table, a formula, and a graph. They make tables using given growth rate information and write direct formulas to describe each growth model.

About the Mathematics Each of the growth models—table, graph, and formula—has its advantages and disadvantages. Students should be able to move flexibly between the three representations.

When a graph is chosen to represent a relationship, the scales of the horizontal and vertical axes should be such that the necessary information can be easily read from the graph. This often means that the scale on the horizontal axis differs from that on the vertical axis.

Planning Students may work on problems **3–5** in small groups. Problem **4** may be assigned as homework. As you discuss these problems, make sure students understand all three representations (table, graph, and formula) and how they are related.

Comments about the Problems

3. a. Make sure students choose an appropriate scale for the vertical axis.

b. The prediction of a value beyond the graph is known as an *extrapolation.*

d. Direct formulas were introduced in *Graphing Equations.* Because the scales on the two axes are different, students should *not* use a protractor to measure the angle of the line and use the tangent of this angle as the slope.

4. Homework This problem may be assigned as homework. It is similar to problem **3.**

Extension Have interested students design and illustrate a working model that shows stepwise linear growth. Encourage students to be creative. Some students may design a sliding device that shows the steps in the growth pattern. Others may even think to make a "flip book" similar to those used to make animated cartoons.

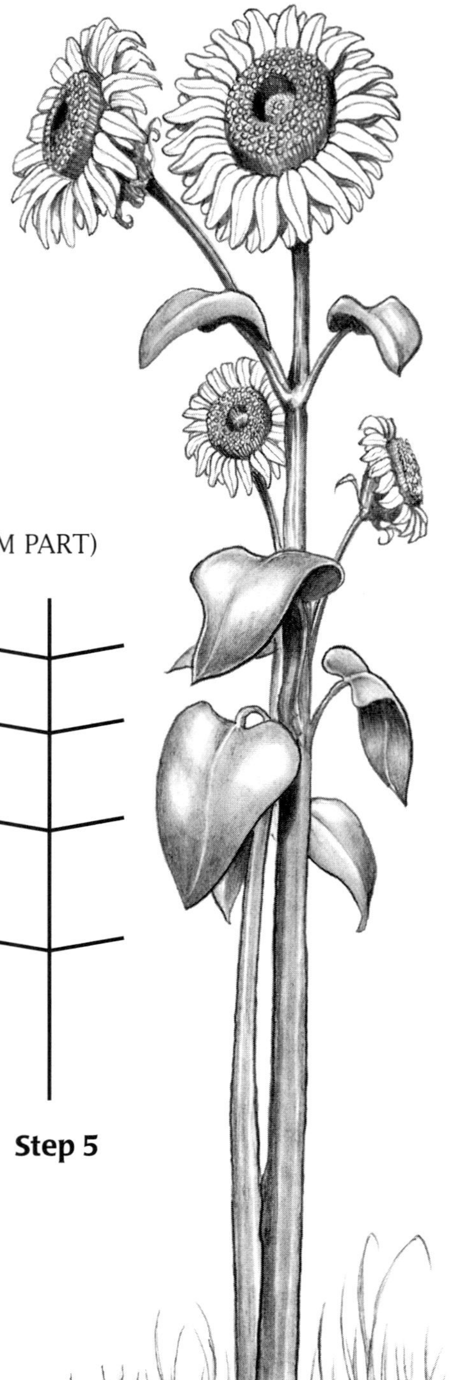

Changing Growth

One possibility for improving our plant growth model is to make each additional stem segment shorter than the previous stem segment. The following formula can be used to describe this type of plant growth:

LENGTH OF NEXT STEM PART = 0.8 × (LENGTH OF CURRENT STEM PART)

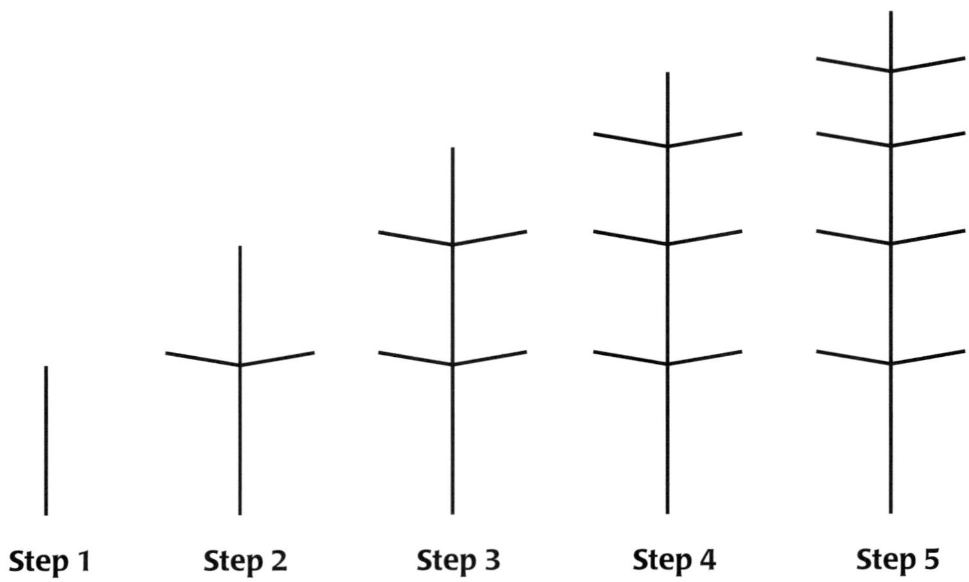

6. Verify that the plant model shown above follows the formula for the lengths of the stem parts.

6. The measured heights are very close to the heights predicted by the formula, so the model follows the formula. Explanations may vary. Some students may measure the lengths of the stem parts and compare those lengths to the corresponding lengths obtained by using the formula, as shown in the table below.

	Length of the Stem Part (in mm) Found by Measuring	Length of the Stem Part (in mm) Found by Using the Formula
Step 1	20	
Step 2	16	16
Step 3	13	12.8
Step 4	10	10.24
Step 5	8	8.2

Overview Students investigate a growth process that is not constant. They examine a model of a growth pattern to verify that it follows the given NEXT-CURRENT formula.

About the Mathematics In constant or linear growth situations, the same amount is added to the starting amount during each time interval. Thus, linear growth patterns and their formulas involve addition.

On this page, exponential growth, which was informally introduced in the unit *Ups and Downs,* is revisited. In exponential growth situations, the amount of growth in one time interval is dependent on that of the previous time interval. Thus, exponential growth patterns and their formulas are based on a multiplicative relationship. NEXT-CURRENT formulas (or recursive formulas) were introduced in *Building Formulas* and studied more extensively in *Ups and Downs.*

Planning Make sure that students understand the notation used in the NEXT-CURRENT formula on this page. Let students complete problem **6** before you discuss this new type of growth pattern. Do not use the term *exponential growth* at this point in the unit. That concept will be formally defined in Section C. Students may work on problem **6** in small groups.

Comments about the Problems

6. Encourage students to make a table to verify the formula. The plant model on Student Book page 11 does not have measurements exactly matching those computed with the given formula. Students may also make minor measurement errors. Make sure that students understand that they should look at the overall picture here and not focus too much on exact measurements.

On **Student Activity Sheet 2,** a plant is shown that follows the formula:

LENGTH OF NEXT STEM PART = 0.8 × (LENGTH OF CURRENT STEM PART)

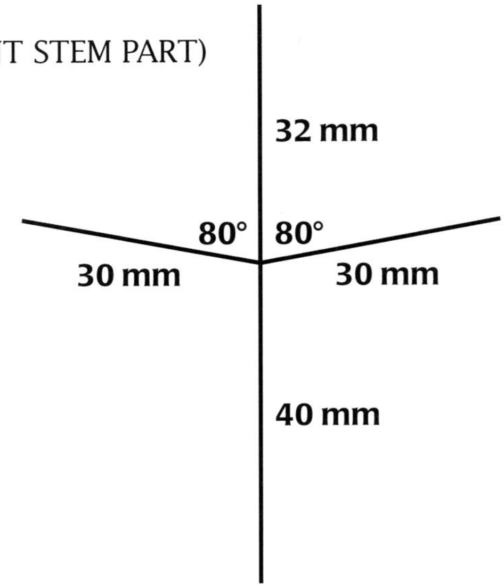

The length of the first stem part is 40 millimeters. The lengths of the branches are all 30 millimeters, and the branches grow at an 80° angle to the stem.

7. Continue the plant model that is shown on **Student Activity Sheet 2** until it has 11 stem parts. Be sure that your measurements are accurate and labeled.

8. Think of your plant as growing one step each day. The first step (the first stem length) is day 0.

a. Copy the following table into your notebook and record the length of the additional stem part for each day. (*Hint:* You can use the formula given above to find the lengths.)

Day	0	1	2	3	4	5	6	7	8	9	10
Length of Added Stem Part (in mm)	40	32									

b. On graph paper, graph the data from your table.

c. Copy the following table into your notebook and record the daily heights of the plant. Make a graph showing the daily heights. Describe the kind of growth displayed in this graph.

Day	0	1	2	3	4	5	6	7	8	9	10
Height (in mm)	40	72									

d. Is the growth of this plant linear? Why or why not?

9. Design your own rule that changes the lengths of the branches with each step. The new plant should look natural. Draw a picture of this plant.

7. The plant model should have 11 stem parts and 10 leaf branches. The stem parts should have lengths as recorded in the table given for the solution to problem **8a** below.

8. a.

Day	0	1	2	3	4	5
Length of Added Stem Part (in mm)	40	32	25.6	20.5	16.4	13.1

Day	6	7	8	9	10
Length of Added Stem Part (in mm)	10.5	8.4	6.7	5.4	4.3

Note: The stem part measurements in the table above have been rounded to the nearest tenth of a millimeter.

b.

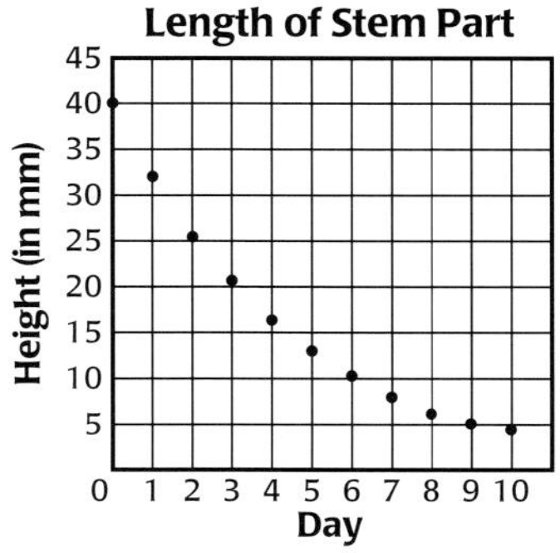

c.

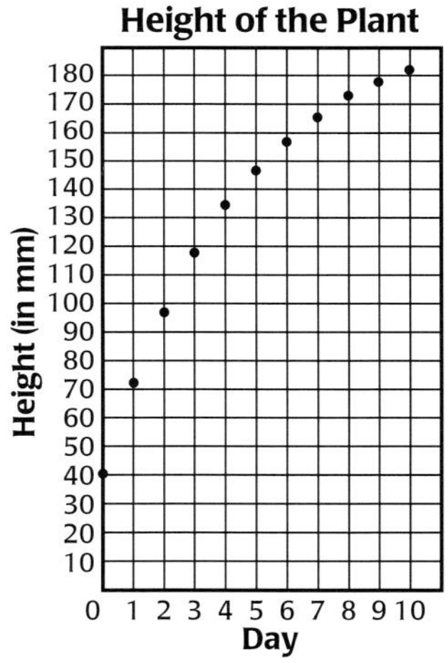

Day	Height (in cm)
0	40
1	72
2	97.6
3	118.1
4	134.5
5	147.6
6	158.1
7	166.5
8	173.2
9	178.6
10	182.9

d. No. Explanations will vary. Sample explanation:

The plant growth pattern is not regular, so its graph is not a straight line. The height of the plant increases by a smaller amount each day. The height continues to increase, but the growth rate decreases.

9. Answers and drawings will vary. Sample response:

The length of each of the first two branches is 60 mm. Each day, the length of each of the next two branches is half the length of the previous branches.

Materials Student Activity Sheet 2 (one per student); graph paper (one sheet per student)

Overview Students are introduced to a new plant model. They use the given formula to continue drawing the model. Students then complete a table, construct a graph, and write a formula that represents this growth pattern.

About the Mathematics The formula that describes the growth process is given. With this formula, the length of each new stem part can be calculated. These values can be recorded in a table and then used to make a graph.

Planning You may want to discuss the text and problems on Student Book page 12. Students may work on problems **7–9** individually. These problems may be assigned as homework. Problem **9** may be used as an informal assessment.

Comments about the Problems

7–9. Homework These problems may be assigned as homework. *Note:* The angle of 80° and the actual length of the branches in the drawing on Student Book page 12 are not relevant for the investigations on this page.

8. Some students may measure the lengths of the stem parts in their drawings and use those measurements for the table. You may want to tell students to use the formula to calculate values for the table.

9. Informal Assessment This problem assesses student's ability to recognize and describe growth patterns presented in pictures and to interpret and create recursive formulas for exponential growth.

Extension Have students use the data from the table and graph in problem **8** to write a direct formula for the length of the added stem part in terms of the number of days that have passed. [The direct formula is $h = 40 \times 0.8^t$, where h = height and t = the number of days elapsed.]

Variation

Your plant model is growing more like a real plant, but there is still a problem with it—all of the plants drawn from this same set of rules are identical. In order for your model to be realistic, there should be some variation among the plants. One way to introduce variation is to include *chance* in your model.

Suppose that the length of each stem part varies between 18 millimeters and 22 millimeters. One way to introduce chance is to use a number cube along with the table on the right to determine the length of a stem part.

Number Cube Outcome	Length of Stem Part
1	18 mm
2	19 mm
3	20 mm
4	20 mm
5	21 mm
6	22 mm

10. Using a number cube and the table, determine and draw several steps of a new plant. How does your plant compare with those drawn by others in class?

Comparing Two Models

Plant A starts with a stem length of 64 millimeters and grows according to the following formula:

LENGTH OF NEXT STEM PART = 0.75 × (LENGTH OF CURRENT STEM PART)

Plant B starts with a stem length of 40 millimeters and grows according to the following formula:

LENGTH OF NEXT STEM PART = LENGTH OF CURRENT STEM PART + 40 millimeters

11. Suppose each step takes one day. When will plant B be taller than plant A? Explain.

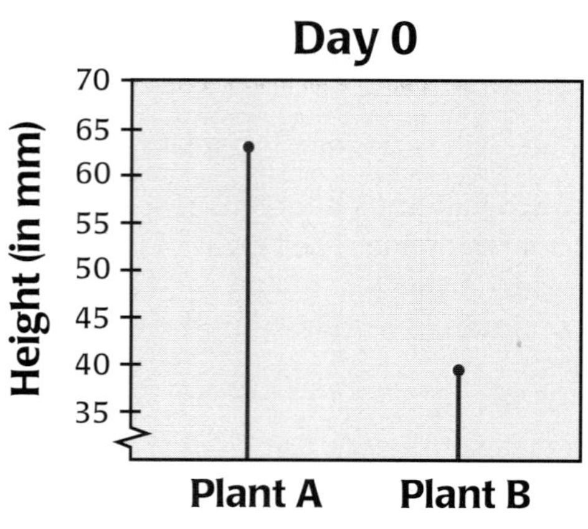

10. Answers will vary, depending on the results students obtain using the number cube and the number of steps drawn.

11. Plant B will be taller than plant A at step 4, or on the fourth day. Explanations will vary. Sample explanation:

I used the growth formulas for plants A and B to compute the heights of the plants for each day and recorded the heights in the table below. The table shows that plant B will be taller than plant A on day 4.

	Plant A	Plant B
Start	64 mm	40 mm
Day 1	112 mm	80 mm
Day 2	148 mm	120 mm
Day 3	175 mm	160 mm
Day 4	195.25 mm	200 mm

Materials graph paper (one sheet per student); number cubes (one per group)

Overview The chance variable is introduced to make the growth model of the plants more realistic. Students compare constant growth to changing growth in a concrete example.

About the Mathematics Chance is often used to make a model more realistic because not all factors that determine growth can be taken into account in a mathematical formula.

Eventually, exponential growth (or changing growth as it is called in this section) becomes faster than linear growth (or constant growth) when the growth factor (the multiplication number) is larger than 1. In the examples in this section, the growth factor is smaller than 1; therefore, the linear growth patterns eventually grow at a faster rate than the exponential growth patterns.

Planning Students may work on problem **10** in small groups. Problem **10** is optional. If time is a concern, you may omit this problem or assign it as homework. Students may work on problem **11** individually. Problem **11** can be used as an informal assessment.

Comments about the Problems

10. This problem is optional. Students should notice slight differences in the steps of the new plant and differences between the plants drawn by fellow classmates. The average length is probably linear.

11. Informal Assessment This problem assesses students' ability to interpret and create graphs and tables representing linear and exponential growth processes and to interpret and create direct and recursive formulas for linear and exponential growth. It also assesses their ability to understand the differences between recursive and direct formulas; to understand the similarities and differences between linear and exponential growth; and to model a growth situation by choosing an appropriate formula and justify the choice.

Extension Have students compare the following two growth processes by graphing each equation and finding the point of intersection.
NEXT = CURRENT + 0.5; start at 2 cm
NEXT = CURRENT × 0.5; start at 24 cm

Summary

To find a *mathematical model* for plant growth, you started with a few rules and then compared the results to what you already knew about plant growth. You began with some very simple simulations and animations and gradually refined the rules.

You worked with NEXT-CURRENT and direct formulas for two types of plant growth:

- *Constant growth*—the length of the next stem part is equal to the length of the current stem part.
- *Changing growth*—the length of the next stem part is equal to the *growth factor* times the length of the current stem part.

You discovered that the graphs for constant growth and changing growth are also different.

Finally, you introduced variability to the growth of a plant by using chance to determine each additional stem length.

Summary Questions

12. How does each formula or model below show constant plant growth?

a. a graph

b. a NEXT-CURRENT formula

c. a table

d. a direct formula

Shown on the right is a picture of a plant created by a computer.

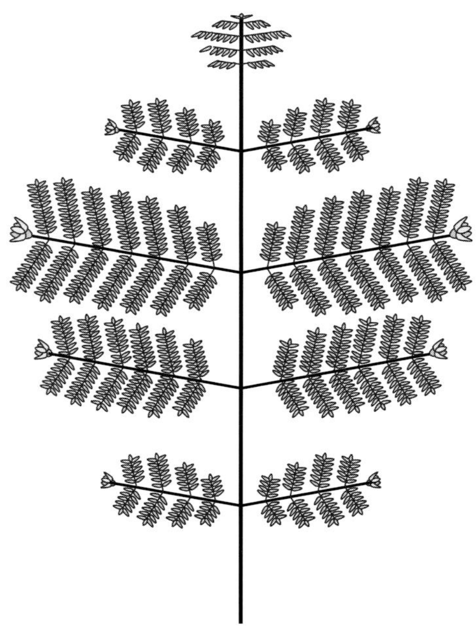

13. If you needed to give a computer instructions for drawing this picture, what directions would you give?

12. **a.** A graph of constant plant growth is always a straight line.

    b. The same amount is always added to the CURRENT amount.

    c. In a table showing data for constant plant growth, the differences between adjacent values in the table are the same.

    d. A direct formula would be height = start + (days × growth per day)

13. Answers will vary. Sample answer:

    Each stem part is 15 millimeters long. Between the stem parts are two symmetrical branches that angle upward 10° above horizontal. The branches in the first pair are each 18 mm long. The branches in the second pair are 27 mm long. The branches in the third pair are 30 mm long. The branches in the fourth pair are the same length as those in the first pair. On top of the fifth stem part are four tiny pairs of symmetrical branches that have the same size relationships to each other as the larger branches do. The branches in the first pair are 3 mm long.

Materials graph paper, optional (one sheet per student)

Overview Students reflect on the models that they used in this section to describe the growth of plants. They explain how each of the representations (graph, table, and formula) show constant plant growth.

About the Mathematics In the Summary text, the term *growth factor* is used for the first time. The growth factor is the number that each current stem part is multiplied by to get the next stem part. For many students, this term may be quite natural. Do not spend too much time explaining this term, since it is formally introduced in Section C.

Planning Students may work individually on problem **12.** This problem may be used as an informal assessment.

After students complete Section B, you may assign appropriate activities from the Try This! section, located on pages 43–46 of the *Growth* Student Book, for homework.

Comments about the Problems

12. Informal Assessment This problem assesses students' ability to recognize and describe growth patterns presented in pictures, tables, and graphs; to interpret and create graphs and tables representing linear growth processes; and to identify common traits within linear formulas.

To get a better impression of students' understanding, you may ask students to come up with an example for each of the four representations.

Writing Opportunity You may have students write their responses to problem **13** in their journals.

SECTION C. EQUATIONS FOR GROWTH

Work Students Do

Students continue to explore the concept of growth. Quadratic growth is used by the students to model the areas of leaves with different heights. Students use cubic growth to model the relationship between the food consumption of a snail and its weight. They explore some of the unique properties of these two models through investigation by using formulas, graphs, and tables.

Goals

Students will:

- recognize and describe growth patterns presented in pictures, tables, and graphs;*
- interpret and create graphs and tables representing quadratic and cubic growth processes;
- interpret and create formulas for quadratic and cubic growth;
- identify the common traits within each type of formula: quadratic and cubic;*
- understand the similarities and differences between quadratic growth and cubic growth;
- understand how changing a formula affects the graph and vice versa;
- model a growth situation by choosing an appropriate formula and justify the choice;
- recognize that a mathematical model usually does not fit a situation perfectly;*
- recognize that growth processes can be simulated with a simplified mathematical model.*

** These goals are assessed in other sections of the unit.*

Pacing

- approximately four or five 45-minute class sessions

Vocabulary

- cubic equation
- exponent
- quadratic equation
- table of first (and second) differences

About the Mathematics

A formula of the form $y = ax^2 + c$, where a and c are constants, is quadratic; the largest exponent is 2. The graph of a quadratic formula is symmetric, and the basic shape is that of the letter U or an upside-down U.

A formula of the form $y = ax^3 + c$, where a and c are constants, is cubic; the largest exponent is 3. A cubic growth process grows faster than a quadratic process.

The values of a and c control the position, width, and direction of the curve; by changing a, the graph becomes more or less steep, while changing c results in vertical shifts of the graph.

Area growth is described with a quadratic formula, and the growth of volume is described with a cubic formula.

Materials

- Student Activity Sheets 3–7, pages 119–123 of the Teacher Guide (one of each per student)
- graph paper, pages 41, 47, 49, and 53 of the Teacher Guide (several sheets per student)
- graphing calculators (or other graphing utility) pages 47, 49, 53, and 61 of the Teacher Guide (one per student)
- centimeter rulers with millimeter scale, pages 51, 53, and 55 of the Teacher Guide (one per student)
- blank transparency, page 49 of the Teacher Guide, optional (one per class)
- transparency of a grid with a graph of an equation on it, page 49 of the Teacher Guide, optional (one per class)
- overhead projector, pages 49 and 53 of the Teacher Guide, optional (one per class)

Planning Instruction

You may introduce this section by discussing the limitations of the growth models explored so far. The linear model predicts unending constant growth; this is not very likely. This section introduces two alternatives: quadratic growth and cubic growth models.

You may need to revisit quadratic equations and area and volume formulas as they have been introduced in earlier *Mathematics in Context* units. Graphing calculators are used in this section. However, if no graphing calculators are available, you can ask students to draw the graphs on graph paper.

Students may work on problems 1 and 2 in small groups or as whole-class activities. They can complete problems 3, 11–14, 17, and 19 in small groups. Students may solve problems 4–9, 15, and 16 individually, in pairs, or in small groups. They may complete problems 10, 18, 20, and 21 individually.

There are no optional problems in this section.

Homework

Problems 9 (page 44 of the Teacher Guide), 16 (page 52 of the Teacher Guide), and 18 (page 56 of the Teacher Guide) can be used for homework. The Writing Opportunity (page 59 of the Teacher Guide) and the Extensions (pages 53 and 59 of the Teacher Guide) can also be assigned as homework. After students complete Section C, you may assign appropriate activities from the Try This! section, located on pages 43–46 of the *Growth* Student Book. The Try This! activities reinforce the key mathematical concepts introduced in this section.

Planning Assessment

- Problem 12 can be used to informally assess students' ability to understand how changing a formula affects the graph and vice versa.
- Problem 18 can be used to informally assess students' ability to interpret graphs representing cubic growth processes.
- Problem 20 can be used to informally assess students' ability to model a growth situation by choosing an appropriate formula and justify the choice, to understand how changing a formula affects the graph and vice versa, to interpret and create a formula for quadratic growth, and to interpret a graph of quadratic growth.
- Problem 21 can be used to informally assess students' ability to understand the similarities and differences between quadratic growth and cubic growth.

C. EQUATIONS FOR GROWTH

Leaf Area

The main function of leaves is to create food for the entire plant. Each leaf absorbs light energy and uses it to decompose the water in the leaf into its elements—hydrogen and oxygen. The oxygen is released into the atmosphere. The hydrogen is combined with carbon dioxide from the atmosphere to create sugars that feed the plant. This process is called photosynthesis.

Here is a small collection of leaves.

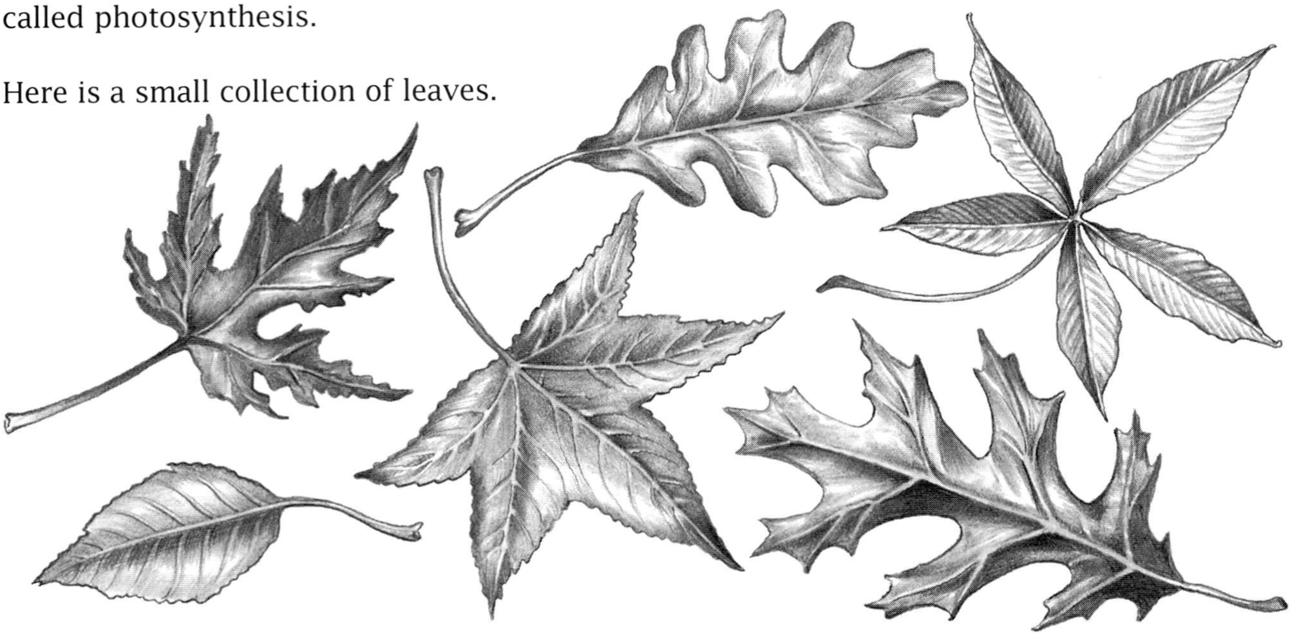

1. **a.** Why do you think a leaf's ability to manufacture plant food might depend on its surface area?

b. Describe a way to find the surface area of one side of any of the leaves shown above.

2. Shown below are three black poplar leaves. Are these leaves similar figures? Why or why not?

Solutions and Samples
of student work

1. **a.** Answers will vary. Sample response:

   A leaf with a larger surface area can absorb more light energy and carbon dioxide and holds more water within it, all of which are required in the manufacture of food.

   b. Answers will vary. Students may suggest tracing a leaf onto a piece of grid paper and counting the squares within the shape, combining any partial squares.

2. Yes, the leaves are similar; they have the same shape even though they are different sizes.

   The ratios of length to width are the same in all figures.

Hints and Comments

Overview Students explore different leaves and describe how to find their surface areas.

About the Mathematics In the unit *Reallotment,* several strategies were used to determine the surface area of a leaf. The leaf of a black poplar can be quite easily modeled by known mathematical shapes; that is why this leaf is used.

Similar means having the same shape—the same ratios between the dimensions and the same angles—but the dimensions (the measurements) themselves can be different. Similarity is also studied in the unit *Triangles and Beyond.*

Planning Students may work on problems **1** and **2** in small groups. These problems may also be completed as a whole-class activity.

Comments about the Problems

1. **b.** You may remind students of what they did in the unit *Reallotment:* comparing the surface areas of leaves by drawing their outlines on grid paper.

2. You may need to refresh students' knowledge of similarity.

Interdisciplinary Connection In science class, students may investigate in more depth how photosynthesis works.

One way to estimate the surface area of one side of a black poplar leaf is to draw a square around it as shown in the diagram below.

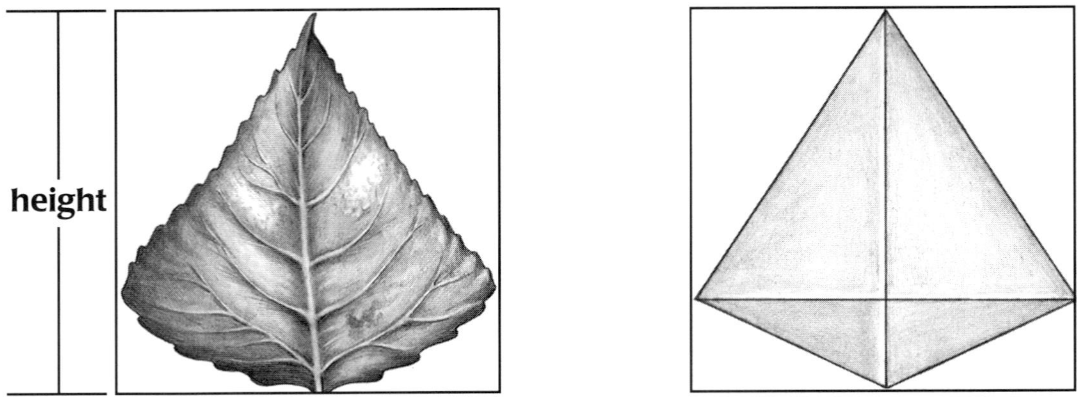

The kite-shaped model on the right covers about the same portion of the square as the actual leaf on the left.

3. a. Approximately what portion of the square does the leaf cover? Explain your reasoning.

b. If you know the height (h) of such a leaf, how can you calculate its area (A)? Write a formula for finding A when you know h. If h is measured in centimeters, what units should be used to express A?

c. The formula that you created in part **b** finds the area of black poplar leaves that are symmetric. Draw a picture of a nonsymmetric leaf for which the formula will still work.

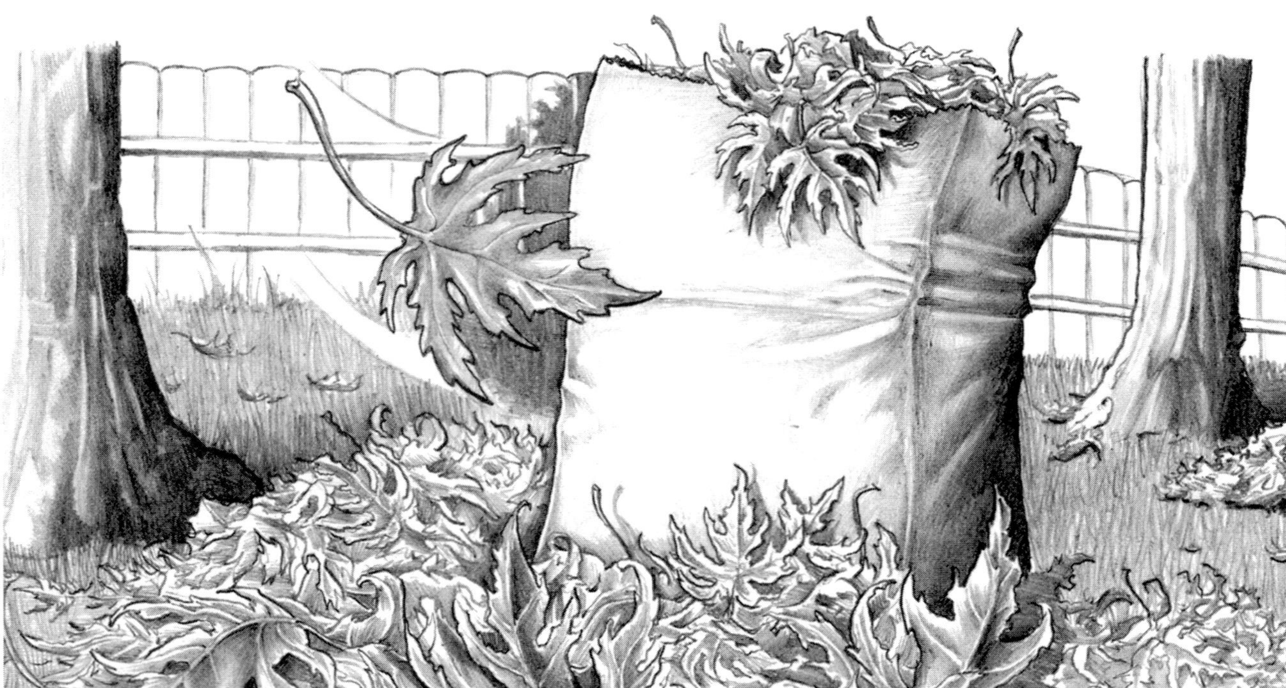

3. **a.** The leaf covers about half of the square. Explanations will vary. Sample responses:

- In the kite-shaped model, if you fold the white part of the figure onto the shaded part, you will have completely covered the shaded part.
- The square can be divided into four smaller rectangles. About half of each rectangle is shaded and half is white.

b. Formulas may vary. Sample formula:

$A = \frac{1}{2}(h^2)$

If the length of the side of the square is h, then the area of that square is h^2. Because the leaf has an area that is half of the area of the square, the formula for the area of the leaf is $A = \frac{1}{2}(h^2)$.

If the height is measured in centimeters, the area should be expressed in square centimeters.

c. Drawings will vary. Sample drawing:

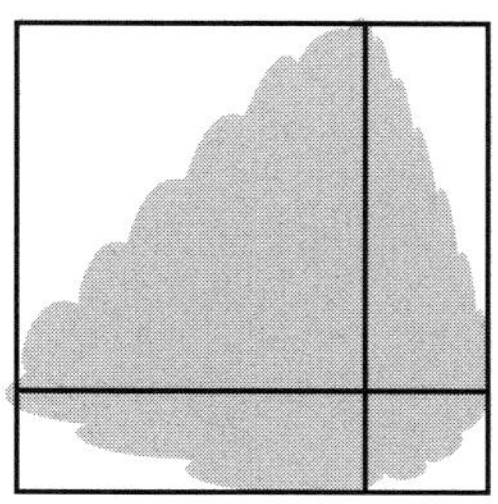

The center of the leaf in this example is shifted to the right.

Materials graph paper (one sheet per student)

Overview Students investigate the surface area of a symmetrical black poplar leaf and create a formula for finding the area based on the height of the leaf. They draw a nonsymmetric leaf that can also be represented by the formula.

About the Mathematics The area of a shape can be found by "reallotting" it, by cutting and pasting the original shape. Another strategy is to divide the given shape into smaller parts of which it is easy to determine the area.

The formula for the area of a square is *length times length* or *length*^{2}. Students should know this formula and notation. Squares and square roots were formally introduced in the unit *Building Formulas.*

Planning You can have students work on problem **3** in small groups. Make sure students know how to work with the formula they find for part **b** before you go on to the next page.

Comments about the Problems

3. **a.** Students should understand that the square can be divided into four pieces such that half of each piece is shaded.

c. Emphasize that students need only draw the outline of the leaf.

Area Differences

Now you have a formula for the area of a black poplar leaf when the height (h) is known:

$$A = \frac{1}{2}(h^2)$$

The table below shows the areas of two black poplar leaves.

Height (in cm)	6	7	8	9	10	11	12
Area (in cm^2)	18	24.5					

4. a. Verify that the areas for heights of 6 centimeters and 7 centimeters are correct in the table.

b. On **Student Activity Sheet 3,** fill in the remaining area values in the table. Describe any patterns that you see.

The diagram below shows the differences between the areas of the first three leaves in the table.

Height (in cm)	**6**		**7**		**8**		**9**		**10**		**11**		**12**
Area (in cm^2)	**18**		**24.5**		**32**		**?**		**?**		**?**		**?**
First Difference		**6.5**		**7.5**		**?**		**?**		**?**		**?**	

5. a. On **Student Activity Sheet 3,** fill in the remaining "first difference" values. Do you see any patterns in the differences?

b. The first "first difference" value (6.5) is plotted on the graph on **Student Activity Sheet 4.** Plot the rest of the differences that you found in part **a** on this graph. Describe your graph.

Solutions and Samples
of student work

4. a. $\frac{1}{2}(6^2) = \frac{1}{2}(36) = 18 \text{ cm}^2$

$\frac{1}{2}(7^2) = \frac{1}{2}(49) = 24.5 \text{ cm}^2$

b.

Height (in cm)	6	7	8	9	10	11	12
Area (in cm^2)	18	24.5	32	40.5	50	60.5	72

Students may note the following patterns:

- For odd heights, the areas end with five-tenths.
- Each area is an increase over the previous area, first by 6.5, then by 7.5, then 8.5, then 9.5, and so on.

5. a.

Height (in cm)	6	7	8	9	10	11	12
Area (in cm^2)	18	24.5	32	40.5	50	60.5	72
First Difference		6.5	7.5	8.5	9.5	10.5	11.5

The differences increase by one each time.

b.

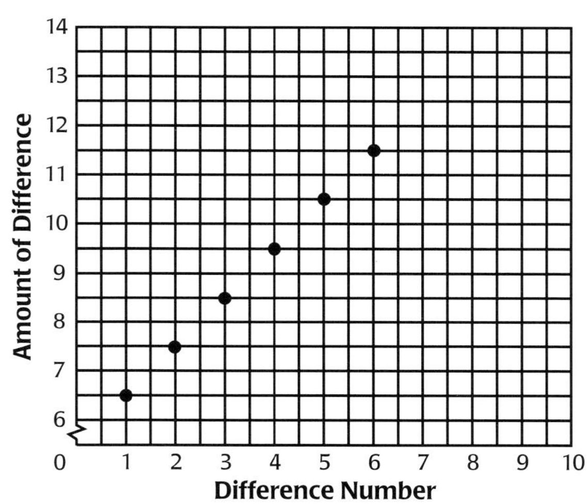

The points lie on a line.

Hints and Comments

Materials Student Activity Sheets 3 and 4 (one of each per student)

Overview Students make a table for the areas of different black poplar leaves and investigate the differences between the areas.

About the Mathematics The study of differences is called differential calculus. It is an intuitive preparation for the more formal mathematics of finding derivatives. In the *Mathematics in Context* curriculum, the study of differences is the endpoint of preparing for calculus. This lays a conceptual foundation for calculus at high school.

Planning Before students begin the problems on this page, you may need to spend some time talking about the idea of differences. Students may work on problems **4** and **5** individually, in pairs, or in small groups.

Comments about the Problems

4. b. Two patterns are apparent just by looking at the numbers in the table: the numbers increase and they are all halves of square numbers.

5. Students worked with similar tables in the unit *Patterns and Figures.*

Height (in cm)	6	7	8	9	10	11	12
Area (in cm²)	18	24.5	32	?	?	?	?
First Difference	6.5	7.5	?	?	?	?	
Second Difference	1	?	?	?	?		

As shown in the above diagram, you can find one more row of differences, called the second differences.

6. **a.** In the diagram on **Student Activity Sheet 3,** finish filling in the row of second differences.

b. What do you notice about the second differences? If the diagram were continued to the right, what would the next two second differences be?

c. How can you use the patterns of the second differences and first differences to find the areas of leaves that have heights of 13 centimeters and 14 centimeters? Continue the diagram on **Student Activity Sheet 3** for these new values.

d. Use the area formula for black poplar leaves ($A = \frac{1}{2}h^2$) to verify your work in part **c.**

7. How does the value of A for a black poplar leaf change when you double the value of h? Use some specific examples to support your answer.

8. If the area of one black poplar leaf is about 65 square centimeters, what is its height? Explain how you found your answer.

9. **a.** On **Student Activity Sheet 5,** fill in the remaining area values in the table using the formula $A = \frac{1}{2}h^2$ and then graph the formula on the provided grid. Why do you think the graph curves upward?

b. Using your graph, estimate the areas of black poplar leaves with the following heights: 5.5 centimeters, 9.3 centimeters, and 11.7 centimeters.

c. Check your answers to part **b** using the formula for the area of a black poplar leaf. Which method do you prefer for finding the area of a black poplar leaf given its height: the graph or the formula? Explain.

Solutions and Samples
of student work

6. a. See the diagram for part **c,** below.

b. The second difference is always 1.

c. You begin with the 1 in the bottom row and work upward.

Height (in cm)	6	7	8	9	10	11	12	13	14
Area (in cm^2)	18	24.5	32	40.5	50	60.5	72	84.5	98
First Difference		6.5	7.5	8.5	9.5	10.5	11.5	12.5	13.5
Second Difference			1	1	1	1	1	1	1

d. $A = \frac{1}{2}(13^2) = 84.5 \text{ cm}^2$; $A = \frac{1}{2}(14^2) = 98 \text{ cm}^2$

7. When h is doubled, A is quadrupled. For example, if $h = 2$ cm, then $A = \frac{1}{2} \times 4 = 2 \text{ cm}^2$. If $h = 4$ cm, then $A = \frac{1}{2} \times 16 = 8 \text{ cm}^2$. Since $8 = 2 \times 4$, the original area of 2 cm^2 quadruples to 8 cm^2 when the height is doubled.

8. Its height is about 11.4 centimeters. Students can use the formula $A = \frac{1}{2}(h^2)$; if $A = 65$, then $h^2 = 130$, and $h = \sqrt{130} \approx 11.4$ cm.

9. a.

Height (in cm)	1	2	3	4	5	6	7	8
Area (in cm^2)	0.5	2	4.5	8	12.5	18	24.5	32

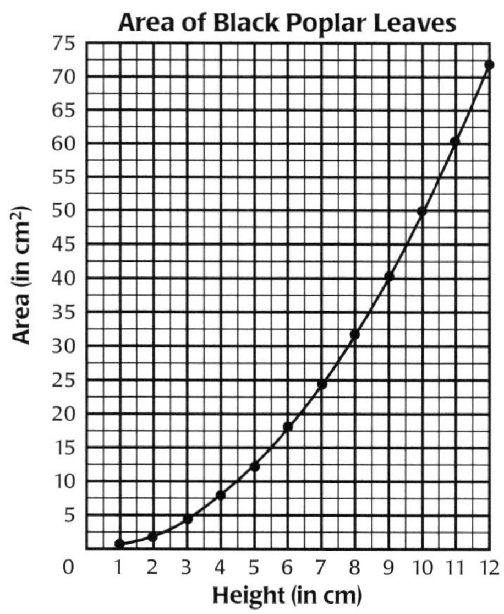

The graph curves upward because the area increases by a larger amount with each change in height.

b–c.

Height (in cm)	Area (in cm^2)	
	Estimated from Graph	Calculated with Formula
5.5	15	15.125
9.3	43	43.245
11	60	60.5

Hints and Comments

Materials Student Activity Sheets 3 and 5 (one of each per student)

Overview Students now investigate the second differences in the areas of the leaves—these are the differences between the differences. Students graph the formula for the area of a leaf and use the graph to estimate the areas of leaves with given heights.

About the Mathematics The formula for the area of a leaf is quadratic. In a quadratic formula, when the *x*-value is doubled, the resulting value quadruples. This can be explained with square diagrams. If the length of a side is doubled, the resulting square is the size of four of the original squares.

Planning After problem **8,** you may want to have an in-depth class discussion about the previous problems. Discuss how the differences can be found and what they mean. Students may work on problems **6–9** individually, in pairs, or in small groups. Problem **9** may be assigned as homework.

Comments about the Problems

7. Make sure students understand the diagram. Problem 7 is critical. At this point students need to realize the common features of quadratic functions.

8. Students should be able to use tables, formulas, and graphs to find one value when given another.

9. Homework This problem may be assigned as homework.

c. Students may have different preferences. It takes less time to estimate the values from the graph, but the formula gives more precise values.

Bringing Math Home After this page, you may ask students to tell family members what they have learned about differences. This may inform parents about the *Mathematics in Context* approach toward developing concepts related to calculus.

Quadratic Equations

You can use a graphing calculator to investigate the graphs of different curves such as $A = \frac{1}{2}h^2$, but your calculator can read only formulas that use X and Y as the variables. Therefore, the formula will have to change to $Y = \frac{1}{2}X^2$ or $Y = 0.5X^2$.

Use a window of $0 < X < 5$ and $0 < Y < 20$ on your graphing calculator for the following problems.

10. **a.** On your graphing calculator, graph $Y = X^2$. Describe the resulting graph.

b. Now graph $Y = 2X^2$ and $Y = 3X^2$. How do these graphs compare with $Y = X^2$?

c. Use either the GRAPH and TRACE features or the TABLE feature of your graphing calculator to determine how the values of Y for $Y = 4X^2$ compare with the values of Y for $Y = 2X^2$. Write a paragraph describing the results of your investigation.

d. Graph $Y = 0.25X^2$. How is this graph different than the previous graphs?

e. Where do you think the graph of $Y = 0.5X^2$ is located in relation to the graphs from parts **a–d?** Check your answer using your graphing calculator.

All of the equations that you have worked with so far have an exponent of 2, as in $y = x^2$. If 2 is the highest exponent in an equation, the equation is called *quadratic.* Here are several quadratic equations:

$y = x^2$

$y = 2x^2$

$y = x^2 + 3$

$y = -x^2 + 3$

$y = x^2 + 6x + 9$

$y = 2x^2 - 4x + 7$

10. a–b.

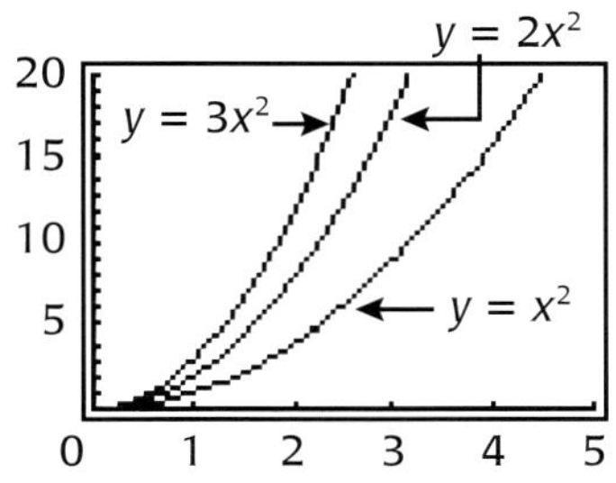

The graphs are curves. The graphs for $y = 2x^2$ and $y = 3x^2$ are steeper than the graph for $y = x^2$.

c. Descriptions will vary. Sample description:

The value of y for the equation $y = 4x^2$ is always twice the y-value for the equation $y = 2x^2$ when the x-value is the same (unless x is 0).

d–e.

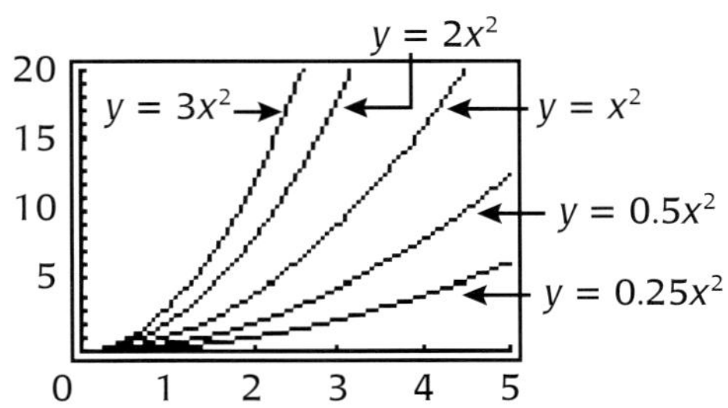

The graph of $y = 0.25x^2$ is one-fourth as steep as the graph of $y = x^2$, one-eighth as steep as $y = 2x^2$, and one-twelfth as steep as $y = 3x^2$.

The graph of $y = 0.5x^2$ is half as steep as the graph of $y = x^2$. It is twice as steep as the graph of $y = 0.25x^2$.

Hints and Comments

Materials graphing calculators (one per student); graph paper, optional (several sheets per student)

Overview Students explore quadratic equations and their graphs.

About the Mathematics Quadratic formulas were introduced in the unit *Building Formulas*, and quadratic patterns were also explored in the unit *Patterns and Figures.*

Graphs can be changed in several ways. When a number is added to (or subtracted) from an equation, the corresponding graph shifts up (or down). When the equation is multiplied by a number larger than one (or smaller than one), the corresponding graph becomes steeper (or less steep). An equation of the form $y = ax^2 + bx + c$ is called *quadratic.* An equation of the form $y = ax + b$ is called *linear.* An equation of the form $ax^3 + bx^2 + cx + d$ is called *cubic* (where a, b, c, and d are all real numbers and not equal to zero).

Planning Students can work on problem **10** individually. You may want students to draw some of the graphs on graph paper in order to develop this skill. You may need to spend some time discussing the mechanics of the graphing calculator for those students unfamiliar with it.

Comments about the Problems

10. If graphing calculators are not available, have students draw the graphs on graph paper.

Extension As an extension to problem **10,** you may use the equations at the bottom of Student Book page 19 to discuss how to change graphs by changing the equations. The graphs can be drawn with or without a graphing calculator. You might also investigate quadratic equations of the form:

$y = (x + 2)^2$

$y = (x - 2)^2$

$y = (x + 2)^2 + 1$

$y = 2(x + 2)^2$

$y = -2(x + 2)^2$

$y = 2(x + 2)^2 + 1$

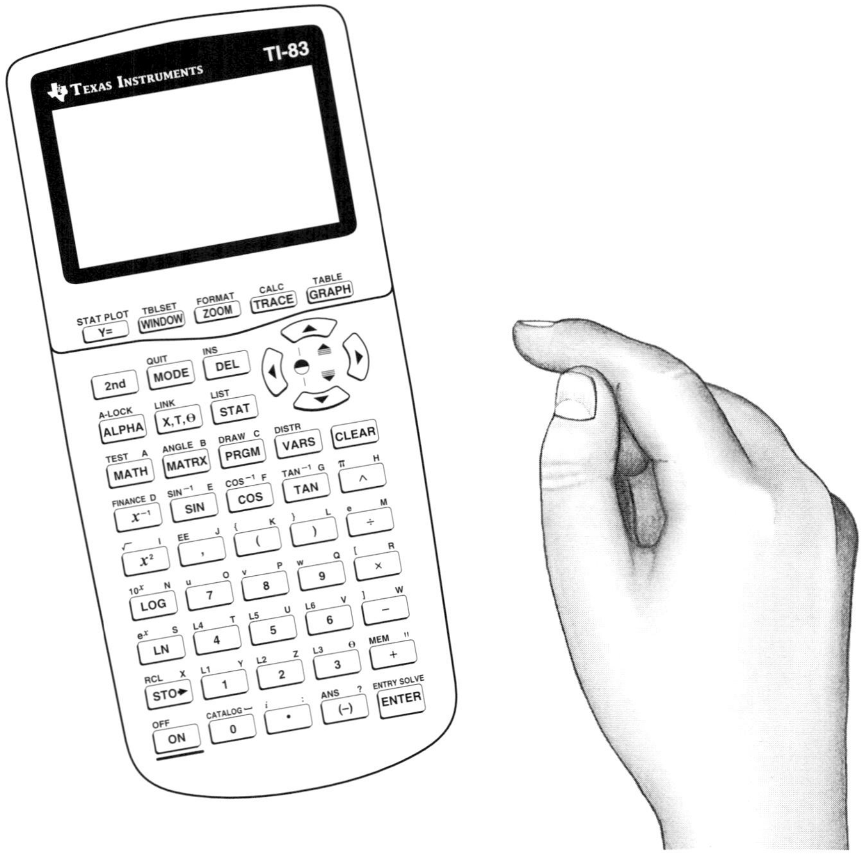

On **Student Activity Sheet 3,** you found a set of first and second differences for the equation $y = \frac{1}{2}x^2$.

11. a. Find a set of differences like you found in problems **5a** and **6a** for at least two other quadratic equations. You can use two of the equations listed at the bottom of page 19 or create two of your own quadratic equations. You might want to make diagrams similar to the one on **Student Activity Sheet 3.**

b. What did you discover about the differences? Can you make any sort of generalization about the differences of quadratic equations?

Using your graphing calculator and a window of $0 < X < 5$ and $0 < Y < 20$, graph $Y = X^2$ and $Y = X^2 + 3$.

12. a. How does adding 3 to $Y = X^2$ affect the graph? (*Hint:* Use the TABLE feature on your graphing calculator.)

b. Something is not quite right with the picture on the calculator screen. What do you notice?

c. Graph $Y = X^2 - 10$ on your graphing calculator. What has happened to the graph? (*Hint:* You might need to adjust the window to get a better picture of this situation.)

Solutions and Samples

of student work

11. a. Answers will vary, depending on the quadratic equations used. Below are the *x* and *y* values and the differences for the quadratic equations given on Student Book page 19.

$y = x^2$

x	6	7	8	9	10	11	12
y	36	49	64	81	100	121	144
difference	13	15	17	19	21	23	

$y = 2x^2$

x	6	7	8	9	10	11	12
y	72	98	128	162	200	242	288
difference	26	30	34	38	42	46	

$y = x^2 + 3$

x	6	7	8	9	10	11	12
y	39	52	67	84	103	124	147
difference	13	15	17	19	21	23	

$y = -x^2 + 3$

x	6	7	8	9	10	11	12
y	−33	−46	−61	−78	−97	−118	−141
difference	13	15	17	19	21	23	

$y = x^2 + 6x + 9$

x	6	7	8	9	10	11	12
y	81	100	121	144	169	196	225
difference	19	21	23	25	27	29	

$y = 2x^2 - 4x + 7$

x	6	7	8	9	10	11	12
y	55	77	103	133	167	205	247
difference	22	26	30	34	38	42	

b. Answers will vary. Students should note that the differences increase by a constant number, which indicates a linear increase.

12. a. It moves the graph of $y = x^2$ up three units. See the table values above.

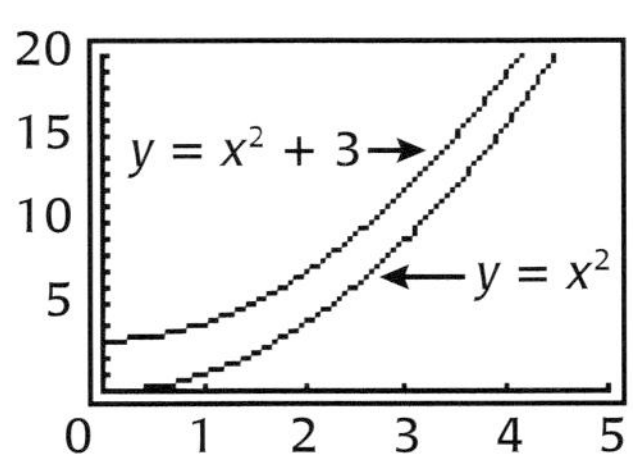

b. Answers will vary. Students may say that the graphs look like they get closer to each other as they move up and to the right and that the graphs should instead always be equidistant. Other students may say that they cannot see the whole shape of the graph, that the part that would represent negative *x*-values is missing.

c.

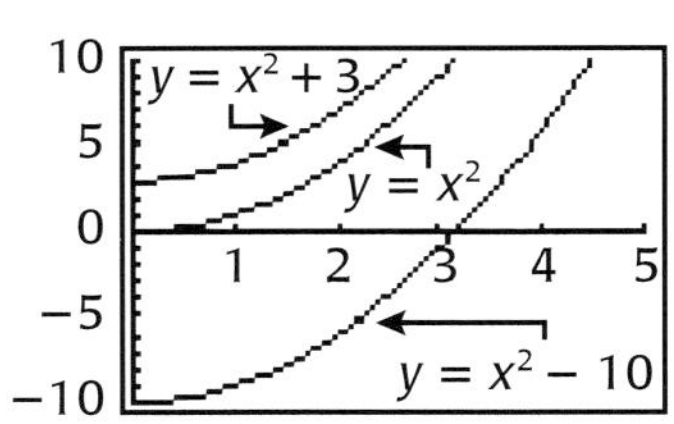

The graph has shifted down by 10.

Hints and Comments

Materials graphing calculators (one per student); graph paper, optional (several sheets per student); blank transparency, optional (one per class); transparency of a grid with a graph of an equation on it, optional (one per class); overhead projector, optional (one per class)

Overview Students continue investigating quadratic equations in relation to the differences.

About the Mathematics For quadratic equations, the first differences are linear and the second differences are constant. In high school mathematics, students will learn why the first derivative of a quadratic relationship is linear.

Planning Students may work on problems **11** and **12** in small groups. After completing the problems on this page, students should understand how they can change graphs by changing the corresponding equations. Problem **12** may be used as an informal assessment.

Comments about the Problems

11. Students should discover that the first differences are linear and that the second differences are constant.

12. Informal Assessment This problem assesses students' ability to understand how changing a formula affects the graph and vice versa.

Extension Make an overhead transparency of a grid with a graph of an equation on it. Then make another transparency with only the graph on it. This allows you to move the copy of the graph over the grid. Then you can ask students what would happen to the equation that represents the graph.

Sizing Up Snails

Some varieties of snails live in freshwater and get their food from the algae on water plants. A group of biologists studied a special type of freshwater snail to determine how the size of the snail affects the amount of food it consumes.

The size of the snail is determined by measuring the length of the shell.

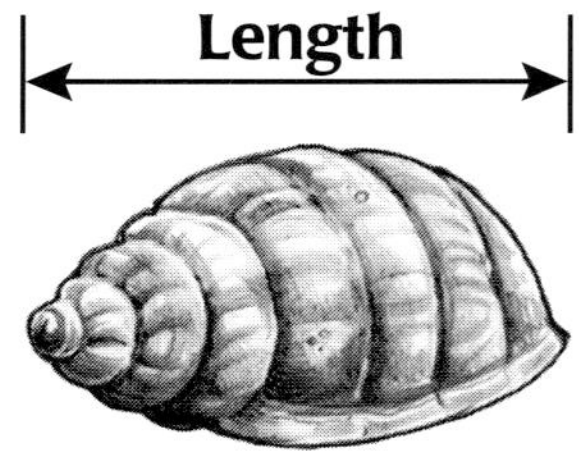

13. What is the length of the snail shell shown on the right in centimeters?

The biologists put a layer of algae on the leaves of test plants in order to find out how much food is eaten by each snail each day. The amount eaten is found by measuring the areas without algae in square centimeters. The graph below shows the collected data.

14. a. In what units do you usually measure food? Do you think the way in which the biologists measured the food is reasonable?

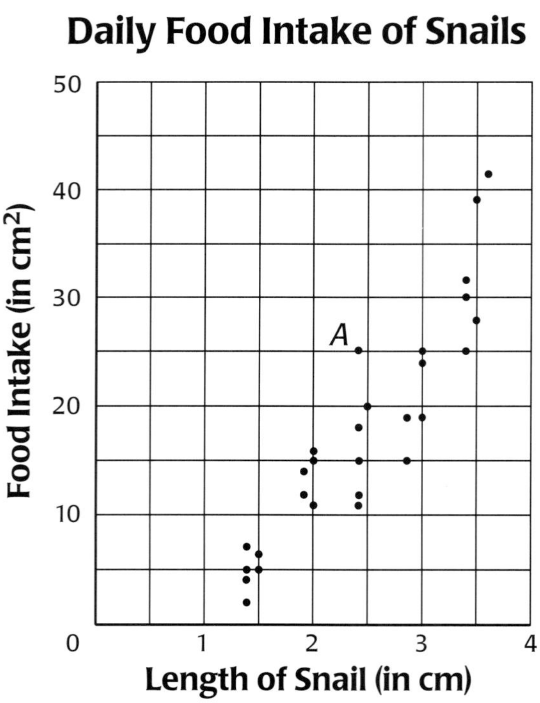

b. What specific information is provided by point *A* in the graph?

c. According to the graph, can two snails that are the same length take in different amounts of food?

d. According to the graph, how is the amount of food consumed by a snail related to its length?

Solutions and Samples
of student work

13. about 3.6 centimeters

14. a. Answers will vary. Students will probably mention the weight units pounds, ounces, and grams. Yes, the unit cm^2 seems reasonable.

b. A snail of almost 2.5 centimeters in length consumed 25 cm^2 of algae in one day.

c. Yes, two snails of the same length can take in different amounts of food. For example, three snails with a length of 2 cm ate 11 cm^2, 15 cm^2, and 16 cm^2 of food.

d. Yes, it is generally true that the bigger the snail, the more it eats.

Hints and Comments

Materials centimeter rulers with millimeter scale (one per student)

Overview Students are introduced to the context of snails, and how the amount they eat is affected by their length.

About the Mathematics The graph on this page is similar to the graph of the plant growth in Section A. A trend can be observed; the longer the snail, the more food it consumes. The next step is to find a model (in this case, a curve) that describes the typical food intake of a typical snail and that can be described by a mathematical rule (an equation or a formula). This process of finding a mathematical model that describes a realistic situation is one of the common activities of mathematicians.

Planning After a brief introduction of the context and a review of the processes of measuring dimensions and reading scatter plots, you may want to have students work on problems **13** and **14** in small groups.

Comments about the Problems

14. This problem is intended to familiarize students with the context and the graph.

To find the behavior of the typical snail, the biologists decided to summarize the data shown on page 21 with a curve. They think that the curve formed by a quadratic equation might fit the data.

It is easiest to explore the data by using a graphing calculator. Listed below are the data that the biologists plotted on the graph on page 21.

15. **a.** Using your graphing calculator and the directions on **Student Activity Sheet 6**, create a scatter plot of the data. What is an appropriate window for viewing these data?

b. Graph $Y = 2X^2$ and $Y = 3X^2$ on your scatter plot. Which equation appears to fit the data better? (Note: X = length, Y = food intake)

c. Using your graphing calculator, try to find a quadratic equation of the form $Y = cX^2$ (where c is a positive number) that fits the data better than either of the two equations that you graphed in part **b.** Explain how you found your answer.

Snail Length (in cm)	Food Intake (in cm^2)	Snail Length (in cm)	Food Intake (in cm^2)
1.4	2	2.4	18
1.4	4	2.4	25
1.4	5	2.5	20
1.4	7	2.8	15
1.5	5	2.8	19
1.5	6.5	3.0	19
1.9	12	3.0	24
1.9	14	3.0	25
2.0	11	3.4	25
2.0	15	3.4	30
2.0	16	3.4	32
2.4	11	3.5	28
2.4	12	3.5	39
2.4	15	3.6	42

Biologists often show their results by drawing a picture. In the drawing shown below, the area of the square represents one day's food intake for a snail whose shell is as long as the pictured shell.

16. **a.** Measure the length (ℓ) of the shell and calculate how much this snail eats in one day using your formula from problem **15c.**

b. How does your answer to part **a** compare with the area of the square that is shown?

c. Draw pictures like the one shown on the right for snails whose shells are 1.5, 2, 2.5, and 3 centimeters long.

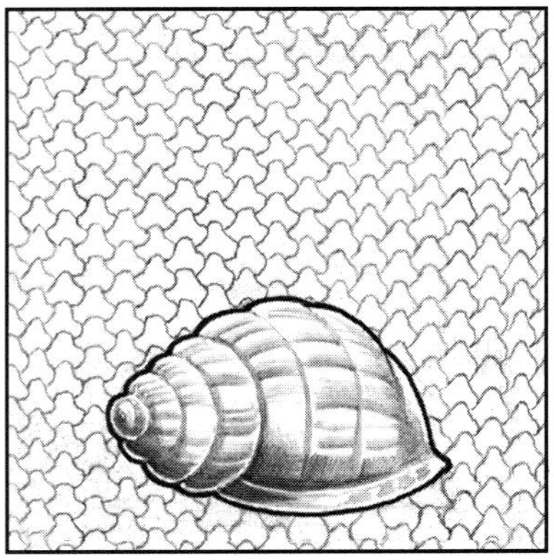

15. a–c.

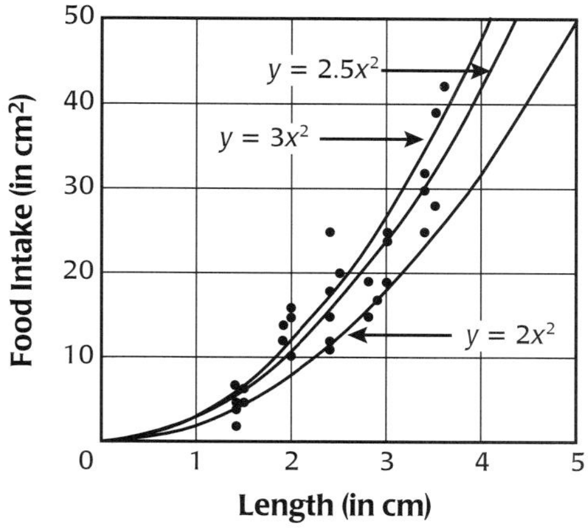

The x-axis represents length, which ranges from 1.4 centimeters to 3.6 centimeters, so a window of $0<x<5$ with an x scale of 0.5 would suffice. The y-axis represents food intake, which ranges from 2 cm² to 42 cm², so a window of $0<y<50$ with a y scale of 5 would suffice.

b. See the diagram above. The equation $y = 3x^2$ seems to fit better than $y = 2x^2$.

c. Answers may vary. A value for c that is between two and three would give a better graph. The graph for $y = 2.5x^2$ (shown in the diagram above) seems to fit well.

16. a. Answers will vary, depending on what value was selected for c in the formula $y = cx^2$. The snail has a length of 3.2 cm. When $y = 2.5x^2$ is used, the food intake is $2.5(3.2)^2 = 25.6$ cm².

b. Answers will vary. The area of the square is around 25 cm². This is fairly close to the answer given for part **a.**

c. Drawings will vary, based on the value for c. If $c = 2.5$, the squares should have sides of approximately 2.4, 3.2, and 4 centimeters, respectively, so that the areas are 5.6 cm², 10.0 cm², and 15.6 cm².

Hints and Comments

Materials Student Activity Sheet 6 (one per student); centimeter rulers with millimeter scale (one per student); graphing calculators (one per student); graph paper, optional (one sheet per student); overhead projector, optional (one per class)

Overview Students plot data and graph equations in order to find an equation that best fits the data. Students also determine the amounts of food that certain snails will eat in a day and draw squares that represent the food intake.

About the Mathematics Curve-fitting can be done by hand, but you can also have a calculator/computer find the equation that fits best. You can specify the kind of equation (linear, quadratic, or other) and the calculator/computer then gives you the best one.

To draw a square for a given area, you need to know the length of the side. Taking the square root of the area gives you the length of the side.

Planning If graphing calculators are not available, you may give students a copy of the graph from the previous page and ask them to draw the graphs for $y = 2x^2$ and $y = 3x^2$ on this same grid.

You may want to revisit squares and square roots before students begin working on problem **16.** This problem may be assigned as homework. Students may work on problems **15** and **16** individually, in pairs, or in small groups.

Comments about the Problems

15. When you have entered these data in a computer or graphing calculator and students have worked on this problem, you may want to graph all kinds of curves on the overhead and have students discuss which is better and why.

16. Homework This problem may be assigned as homework.

Extension A good extension for problem **15** is to have students find equations for the upper and lower curves of the band graph that would include all of the snail data.

A snail's volume can be approximated by multiplying its length times its width times its height. The biologists are also interested in studying a snail's volume (V) in relation to its length (ℓ) only. This relationship will enable them to make only one measurement per snail.

Shown below are a side view and top view of a snail.

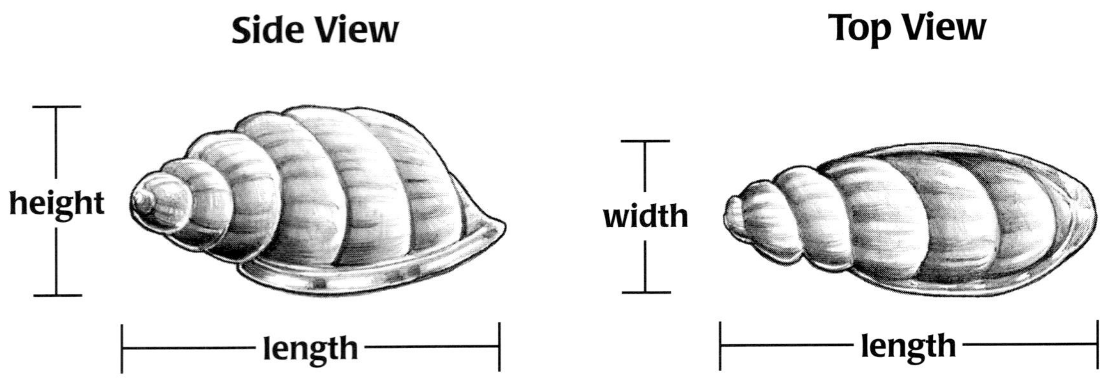

17. a. How does the height of this snail compare to its length?

b. How does the width of this snail compare to its length?

c. What is a possible formula for the volume of this snail based on its length? To find your formula for the volume of a snail, what are some assumptions that you are making about the snail's shape?

The volume of a snail shell is determined by its length, width, and height. Multiplying these three factors results in a formula with an exponent of 3. An equation in which the largest exponent is 3 is called a *cubic equation.*

Solutions and Samples

of student work

17. a. The height is 2.5 cm and the length is 5 cm, so the height is half of the length.

b. The width is 2 cm. The length is 2.5 times the width.

c. Answers will vary. Sample response:

A possible formula is $V = \frac{1}{5}(\ell^3)$. This is based on the following:

$V = \ell \times w \times h$

$w = \frac{2}{5}\ell$

$h = \frac{1}{2}\ell$

This results in the formula:

$V = \ell \times (\frac{2}{5}\ell) \times (\frac{1}{2}\ell)$

$V = \frac{1}{5}(\ell^3)$

When you use the formula, you assume that the snail has the shape of a rectangular prism.

Hints and Comments

Materials centimeter rulers with millimeter scale (one per student)

Overview Students investigate three dimensions of a snail and devise a formula for the volume.

About the Mathematics The volume of an object is often described as length × width × height. This formula, however, only works for rectangular prisms. For other irregular shapes, the formulas for volume are more complex. Therefore, an estimation of volume using length, height, and width is often the easiest. In the unit *Cereal Numbers,* students worked with volumes and ratios explicitly.

While area is described with a quadratic equation, volume is described with a cubic equation. In a quadratic equation, the highest exponent is 2, or in other words, there is a square in the equation. In a cubic equation, the highest exponent is a 3. A square is two-dimensional, and a cube is three-dimensional. This fact may help students remember the meanings of the exponents 2 and 3 and in what contexts the equations are used.

Planning You may want to introduce this page with a brief class discussion about dimensions and how to make comparisons using ratios. Students may work on problem **17** in small groups.

Comments about the Problems

17. Students can look at the ratio of height to length and the ratio of width to length and use these relationships to come up with a formula for the volume based on only the length.

17. c. You may want to spend some time discussing how to simplify formulas and looking at equivalent formulas. Talk with students about how to simplify their equations to get an exponent of 3 for the variable ℓ. For example, $V = 0.2\ell^3$.

Another important measure of a snail is its weight (W). The data points on the graph at the bottom of the page show the relationship between a snail's length and its weight. A computer was used to find the curve that fits the data best. The formula that the computer found for this curve is $W = 0.07\ell^3$, where W is the weight of the snail and ℓ is the length.

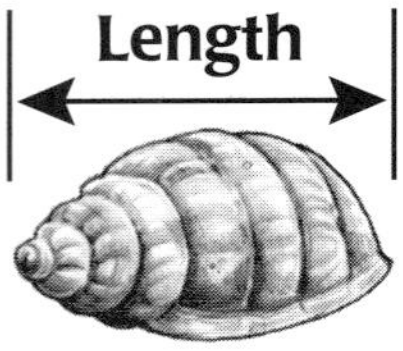

18. a. Use both the graph below and the formula given above to approximate the weight of the snail shown on the right.

b. If you knew the weight of a snail, explain how you could find its length. Write a short paragraph for your explanation. You might want to include some examples of different weights in your explanation.

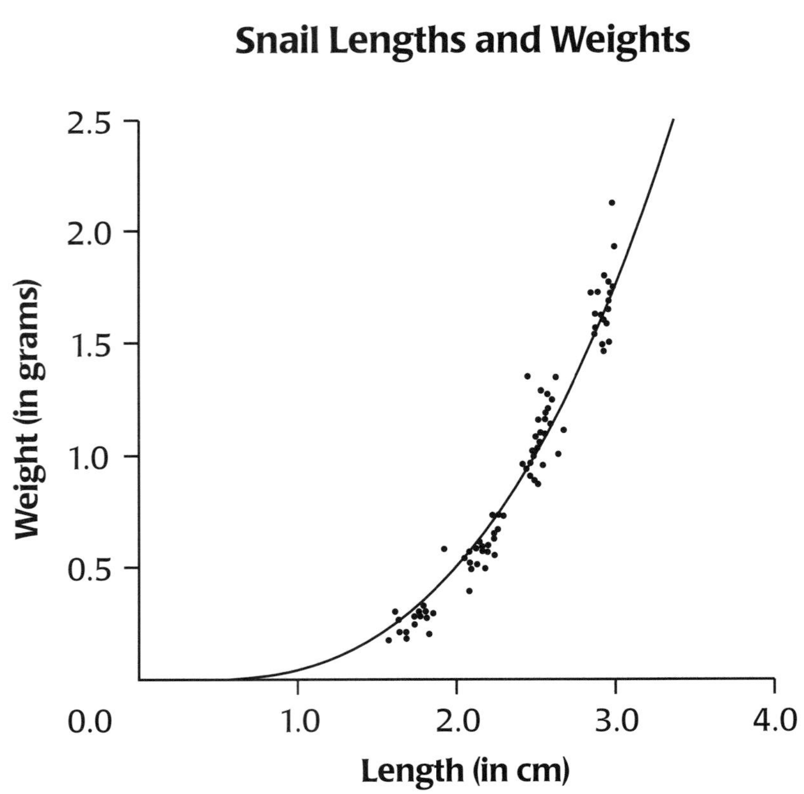

18. a. The length is 2.5 cm. With the formula, this leads to a weight of about 1.1 grams [$W = 0.07\ (2.5)^3$].

Using the curve, you get a weight of about 1 gram; looking at the data points, you find a range between 0.9 and 1.4 grams.

b. Explanations will vary. Sample explanations:

- You could divide the weight by 0.07 and then take the cube root using a calculator. For a weight of 2 grams:
  $2 = 0.07\ (\ell^3)$
  $28.57 \approx \ell^3$
  $\ell \approx 3$ cm
- You can use the graph and round off the values. For example, a weight of 1.5 grams gives a length of about 2.8 centimeters.

Overview Students investigate the relationship between the weight of a snail and its length.

About the Mathematics Students have an opportunity to use both a formula and a graph to approximate the sizes of different snails.

Planning Students can work on problem **18** individually. The problem can be used as an informal assessment and may be assigned as homework.

Comments about the Problems

18. Informal Assessment This problem assesses students' ability to interpret graphs representing cubic growth processes. You may want to assign this problem as homework.

Did You Know? Weight is related to volume. When the density of an object is equal to 1 (as it is for water), then the weight is equal to the volume; volume measured in cubic centimeters gives weight measured in grams.

A Mathematical Investigation

So far in this section, you have worked with quadratic equations such as $y = 0.5x^2$ and cubic equations such as $y = 0.07x^3$. The graphs for these two equations look very similar—they both curve upward. This leads to the question: What is the difference between the growth that is described by a quadratic equation ($y = x^2$) and that described by a cubic equation ($y = x^3$)?

19. a. On the first grid on **Student Activity Sheet 7,** graph $y = x^2$ and $y = x^3$ for values of x between 0 and 5 ($0 \leq x \leq 5$). How do the two graphs compare?

b. On the second grid on **Student Activity Sheet 7,** graph $y = x^2$ and $y = x^3$ for values of x between 0 and 1.5 ($0 \leq x \leq 1.5$). Now how do the two graphs compare?

Summary

The study of growth is the study of change. Growth can be described by a graph or an equation. In this section, you studied the growth of leaves and snails and used two types of equations, quadratic and cubic, to describe the growth.

A quadratic equation is one in which the largest exponent is 2, such as $y = 2x^2 - 5x + 3$. By using your graphing calculator to compare several quadratic equations of the form $y = cx^2$, you discovered that the larger c is, the more quickly the graph grows. For example, $y = 4x^2$ grows more quickly than $y = x^2$.

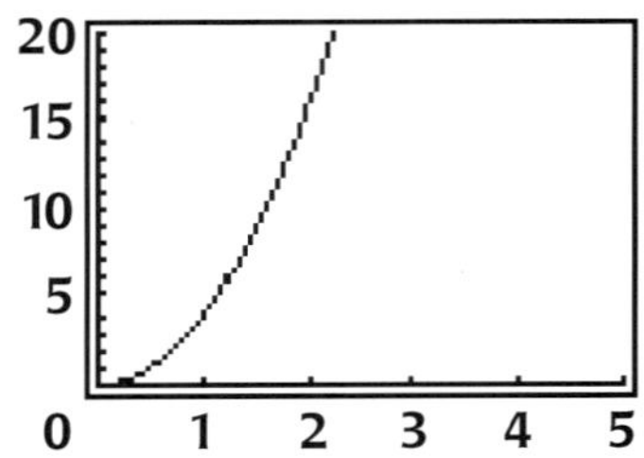

$y = 4x^2$

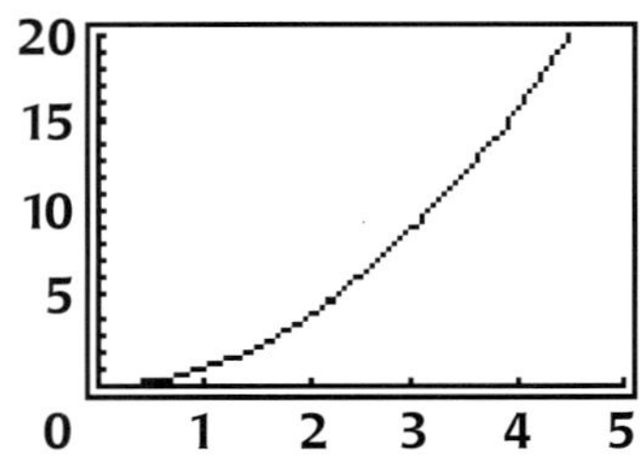

$y = x^2$

Solutions and Samples
of student work

19. a.

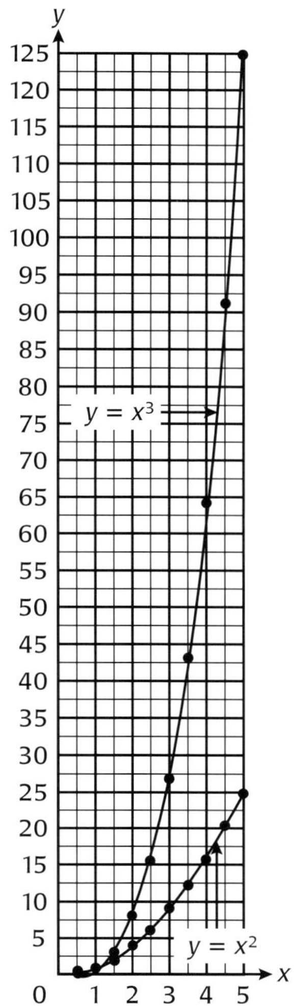

The graph of $y = x^3$ becomes much steeper (goes up faster) than the graph of $y = x^2$.

b.

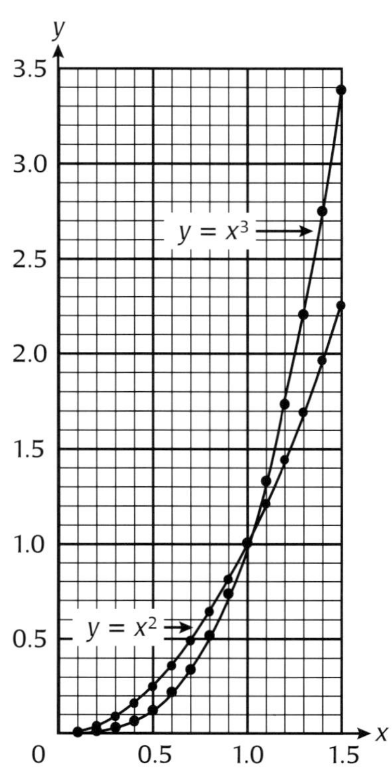

Between the x values of 0 and 1, the graph of $y = x^3$ lies under the graph of $y = x^2$. At (1, 1), the graphs intersect, and when x is greater than 1, $y = x^3$ becomes much steeper than $y = x^2$.

Hints and Comments

Materials Student Activity Sheet 7 (one per student)

Overview Students compare the graphs for $y = x^2$ and $y = x^3$, and they read and discuss the Summary, which reviews the uses for and differences between quadratic and cubic graphs and equations.

About the Mathematics The graph of the equation $y = cx^3$ will eventually rise above the graph of the equation $y = cx^2$ for any value of c. For example, the graph of $y = 100x^2$ is above the graph of $y = 0.5x^3$ for values of x up to 200, but then they intersect and the graph representing cubic growth surpasses the other.

Planning Have students do problem **19** in small groups and discuss their findings before opening up a class discussion of the problem.

Extension 1 Ask students to extend their graphs from problem **19** to show x values to -3.

Extension 2 You may want to challenge students by having them compare the equations $y = 100x^2$ and $y = 0.5x^3$ Have them explain why one rises above the other, although it takes 200 steps.

Writing Opportunity You may ask students to write their answers to problem **19** in their journals.

Summary, continued

By comparing $y = x^2$ and $y = x^2 + 3$ on your graphing calculator, you also discovered that adding 3 to $y = x^2$ moves the graph up by 3.

A cubic equation is similar to a quadratic equation, but the largest exponent of a cubic equation is 3, as in $y = x^3$.

When the value for x is greater than one, a cubic equation grows faster than a quadratic equation.

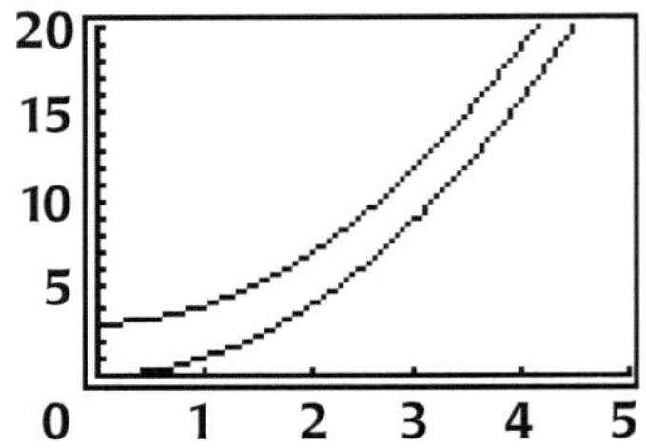

$y = x^2$
$y = x^2 + 3$

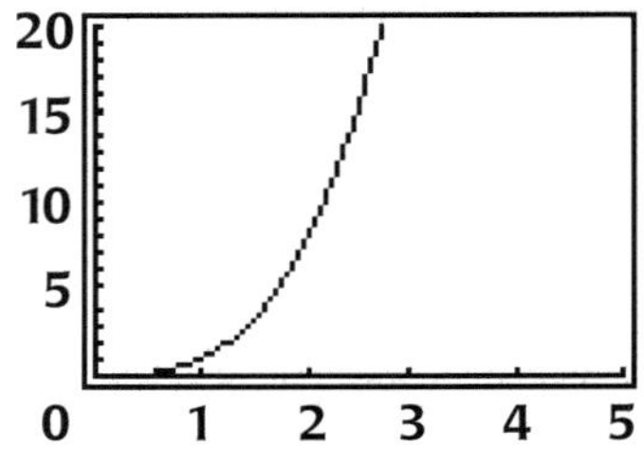

$y = x^3$

Summary Questions

20. The graph shown below represents a growth process. Using your graphing calculator, determine a quadratic equation whose graph approximates the one shown below. Your equation should be of the form $y = cx^2$. Write a few sentences explaining how you found your equation.

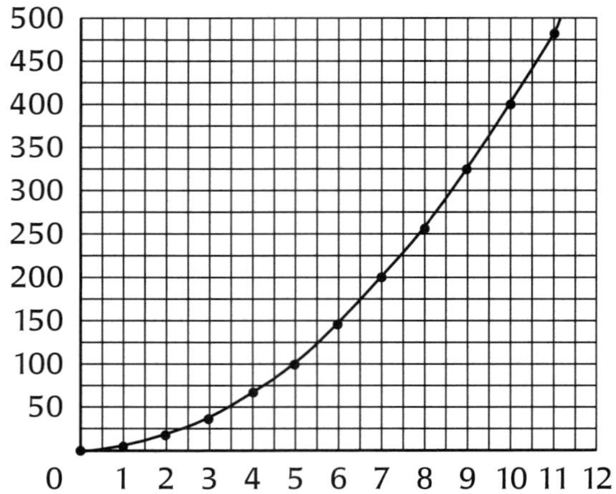

21. Which of the following equations represents the fastest growth? Explain.

a. $y = 3x^2$

b. $y = 0.3x^2$

c. $y = x^3$

d. $y = 3x^3$

Solutions and Samples
of student work

20. Answers and explanations will vary. Sample strategy:

Using a trial-and-error approach, start with $y = x^2$; for $x = 10$, this gives $y = 100$. In the graph, the point is (10, 400), so the equation should be $y = 4x^2$.

21. Equation **d** represents the fastest growth. Explanations will vary. Sample explanation:

Equation **d** represents the fastest growth because it is cubic and grows three times faster than the other cubic equation, $y = x^3$.

Hints and Comments

Materials graphing calculators (one per student)

Overview Students finish reading the section Summary and compare different kinds of graphs and their corresponding equations.

About the Mathematics In this section, students learned the following concepts:

- that adding or subtracting a value to an equation shifts the graph
- that multiplying an equation by a value larger or smaller than 1 makes a graph steeper or less steep
- the differences between quadratic growth and cubic growth
- the meaning and influence of c in the equations $y = cx^2$ and $y = cx^3$

Planning When you summarize this section, address the issues listed in the About the Mathematics section above. Students may work individually on problems **20** and **21.** These problems may be used as informal assessments.

After students complete Section C, you may assign appropriate activities from the Try This! section, located on pages 43–46 of the *Growth* Student Book, for homework.

Comments about the Problems

20. Informal Assessment This problem assesses students' ability to model a growth situation by choosing an appropriate formula and justify the choice, to understand how changing a formula affects the graph and vice versa, to interpret and create a formula for quadratic growth, and to interpret a graph of quadratic growth.

If graphing calculators are not available, you can ask students to draw two graphs (for two values of c).

21. Informal Assessment This problem assesses students' ability to understand the similarities and differences between quadratic growth and cubic growth.

Work Students Do

Students are introduced to explosive growth and to a diagram that models this growth. They explore recursion and construct and use NEXT-CURRENT formulas and direct formulas. Subscripts and arrow language are introduced and used by students to label the steps in a recursive process. With these tools, students explore successive population sizes.

Goals

Students will:

- recognize and describe growth patterns presented in pictures, tables, and graphs;
- interpret and create graphs representing exponential growth;
- interpret and create direct and recursive formulas for linear and exponential growth;
- identify common traits within each type of formula: linear and exponential;
- understand the similarities and differences between linear and exponential growth;
- model a growth situation by choosing an appropriate formula and justify the choice;*
- recognize that growth processes can be simulated with a simplified mathematical model.*

* *These goals are assessed in other sections of the unit.*

Pacing

- approximately two to three 45-minute class sessions

Vocabulary

- arrow language
- explosive growth
- growth factor
- recursive formula
- starting population

About the Mathematics

If growth follows a model in which the value at one point is a multiple of the value at a previous point, this is explosive growth. Another term for this is *exponential growth.* Explosive growth can be described by recursive formulas, which use the current value to find the next value; these formulas are also known as NEXT-CURRENT formulas. The number that the current value is multiplied by to get the next value is called the growth factor. A starting value is needed to start a growth process. The starting value is the time value (t) at the beginning of the growth period, when $t = 0$.

Materials

- Student Activity Sheets 8 and 9, pages 124 and 125 of the Teacher Guide (one of each per student)
- transparency of Student Activity Sheet 8, page 69 of the Teacher Guide, optional (one per class)
- overhead projector, page 69 of the Teacher Guide, optional (one per class)

Planning Instruction

You may wish to introduce this section with a discussion of what would happen if a baby grew by doubling its height every year. Students may be surprised to learn that this growth rate is realistic at first. It does, however, quickly become ridiculous.

Have students work on problems 3 through 6 in one run. You may need to spend some time in class on the "explosive growth diagram" before students start working with it. In this section, students develop a thorough understanding of explosive growth and of the concept of growth factor.

You may choose to have students work in small groups on problems 1–6 and 10–12 and individually on problems 7–9 and 13–16.

There are no optional problems in this section.

Homework

Problems 7 and 8 (page 72 of the Teacher Guide) and problems 13 and 14 (page 76 of the Teacher Guide) can be assigned as homework. In addition, the Extension activities (pages 65, 67, and 79 of the Teacher Guide) may be assigned as homework. After students complete Section D, you may assign appropriate activities from the Try This! section, located on pages 43–46 of the *Growth* Student Book.

Planning Assessment

- Problem 9 can be used to informally assess students' ability to create and interpret graphs for representing exponential growth.
- Problem 15 can be used to informally assess students' ability to recognize and describe growth patterns presented in pictures, tables, and graphs; to interpret and create direct and recursive formulas for linear and exponential growth; to identify common traits within each type of formula: linear and exponential; and to understand the similarities and differences between linear and exponential growth.
- Problem 16 can be used to informally assess students' ability to interpret and create recursive formulas for linear and exponential growth.

D. EXPLOSIVE GROWTH

CaterpillarsCaterpillars

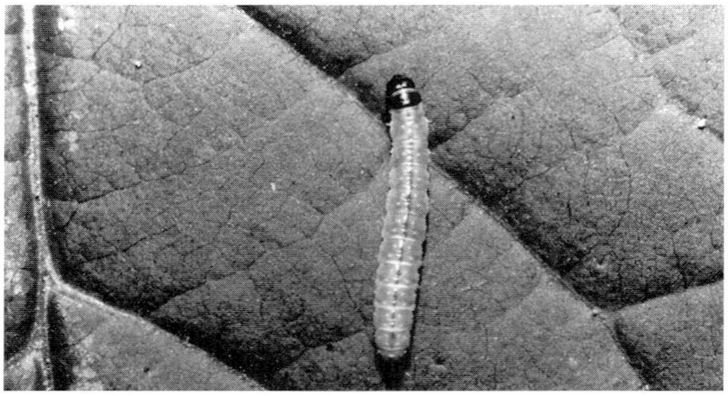
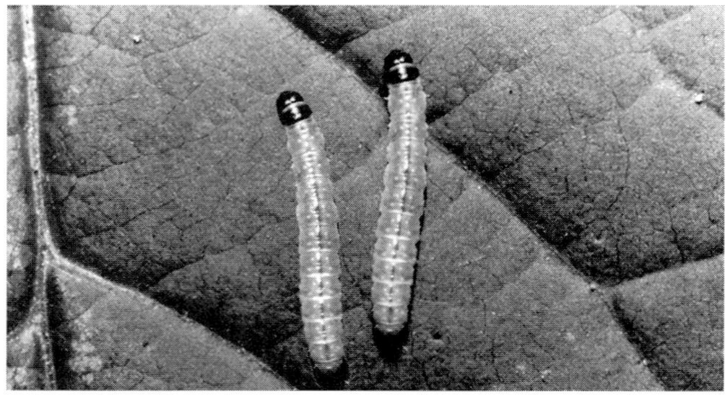
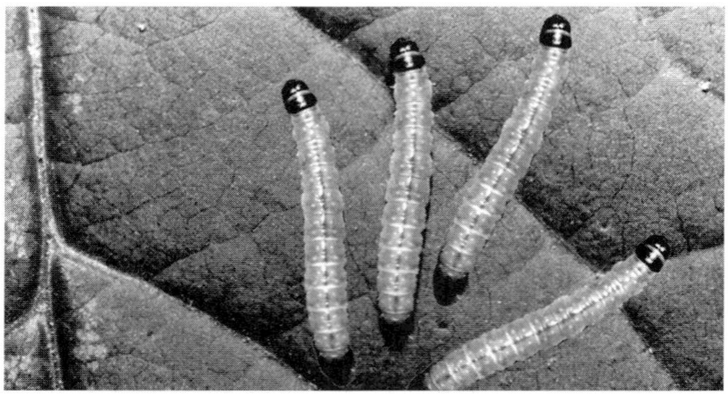
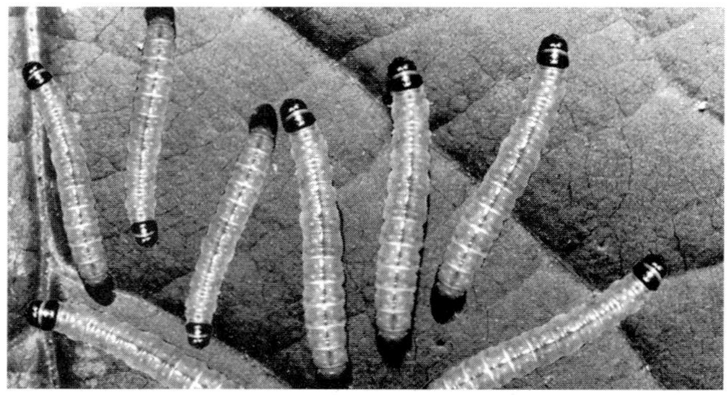

Mr. Jones has had a garden in his yard for many years. This summer he notices twice as many caterpillars on his tomato plants as last summer. He recalls that last summer there were twice as many caterpillars as in the previous summer. Mr. Jones thinks, "If this increase continues, the caterpillars will completely destroy my tomato plants next summer."

Mr. Jones is noticing a rapid growth in the number of caterpillars in his garden. This type of growth can be called *explosive growth.* Explosive growth occurs when an initial period of seemingly slow growth is followed by increasingly fast growth. The rate of growth builds up in a relatively short time.

1. Which of the three graphs shown below best represents explosive growth?

a.

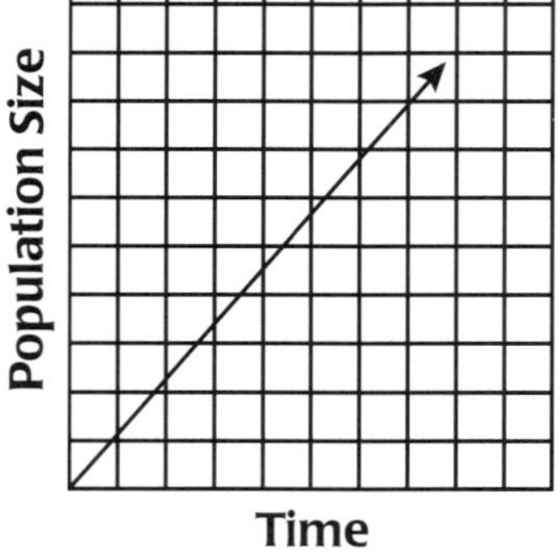

b.

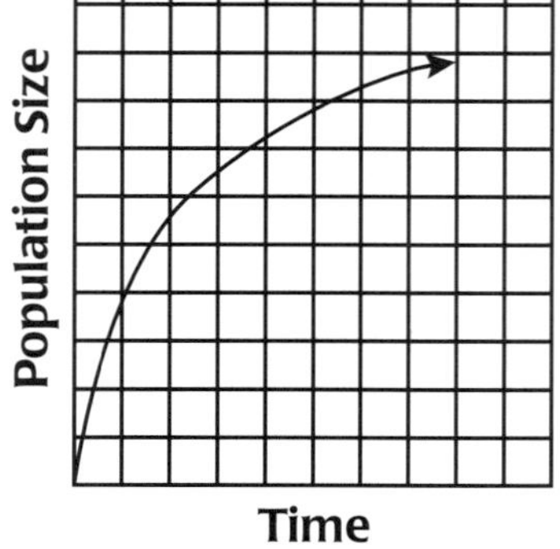

c.

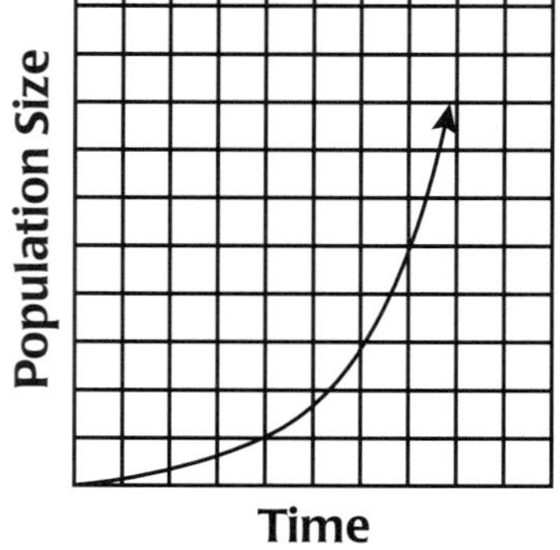

Solutions and Samples

of student work

1. Graph C. Graph A represents constant (linear) growth, and graph B represents a process in which the growth slows down.

Hints and Comments

Overview Students are introduced to the context of a growing caterpillar population. They compare different kinds of growth represented by graphs.

About the Mathematics *Explosive growth* as introduced on this page is the informal term for exponential growth with a growth factor larger than 1. The term *growth factor* is introduced later in this section. The growth can be expressed recursively with the formula

NEXT = CURRENT × growth factor.

In constant growth (as illustrated with graph **a** on Student Book page 27), a constant amount is added each time; in other words,

NEXT = CURRENT + constant.

For decreasing growth (as indicated in graph **b** on Student Book page 27), the growth factor is between 0 and 1. For increasing growth (graph **c**), the growth factor is larger than 1.

Planning You can briefly introduce this section by talking about explosive growth. You may want to refer to the unit *Ups and Downs,* in which students informally studied this kind of growth. Then have students do problem **1** in small groups.

Comments about the Problems

1. Have students describe each graph in words. When you discuss this problem in class, you may want to use the terms *increasing growth* and *decreasing growth.*

Extension Ask students whether they can think of examples of growth that could be represented by each of the three graphs on Student Book page 27.

A caterpillar plague usually builds up over several years. In order to determine whether caterpillar growth is explosive, you need an accurate count of the caterpillar population.

The following graph shows the caterpillar population over a four-year period in Mr. Jones's garden. Year 0 is when Mr. Jones first counted the caterpillars.

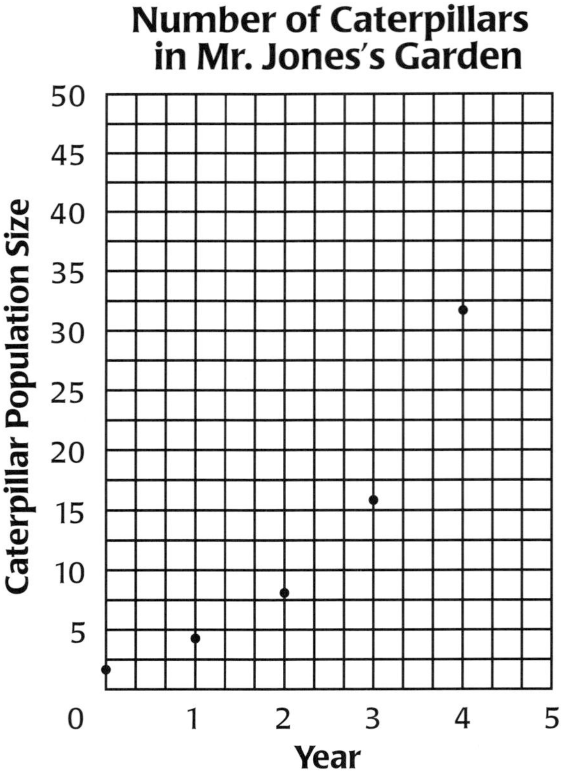

2. a. What is the starting population of caterpillars for this graph?

b. Copy the following table into your notebook and use the graph on the left to fill it in.

Year	0	1	2	3	4
Population					

c. If you draw a curve through the five points, what are you assuming about the growth during the time periods between the points?

Solutions and Samples
of student work

2. a. The starting population is two caterpillars.

b.

Year	0	1	2	3	4
Population	2	4	8	16	32

c. You are assuming that the population grows continuously and not in spurts. You are also assuming that nonintegral values are meaningless within the context of a caterpillar population. For example, you cannot have a caterpillar population of 4.7 caterpillars.

Hints and Comments

Overview Students continue the investigation of the growth of a population of caterpillars.

About the Mathematics On Student Book page 28, a table and a graph are used to investigate growth. The growth studied on this page is different from the quadratic and cubic growth processes studied in the previous section. This can be "proven" by looking at the differences.

Planning Have students work on problem **2** in small groups and then have a brief class discussion about problems **1** and **2.** You may want to discuss the differences between this kind of growth and quadratic and cubic growth. Students need only understand intuitively that explosive growth is different from the growth processes studied so far.

Comments about the Problems

2. Students should use the data in the graph, but do not need to copy the graph. Problem **2c** asks students what it means to draw a curve through the points. Drawing a curve through data points is not always meaningful in the context. In this context, caterpillars are "discrete"; that is, each caterpillar must be counted as a distinct unit, so a population of 24.45 caterpillars, for instance, is not realistic. Students should realize that the mathematical model (the curve) assumes that any number of caterpillars (including fractional amounts) is possible. Recognizing that a curve would misrepresent the data contributes to the development of a critical attitude toward the use of mathematics.

Extension Ask students to come up with a formula to describe the growth of the caterpillar population.

Explosive Growth Display

There is another way to display the explosive caterpillar growth that is occurring in Mr. Jones's garden.

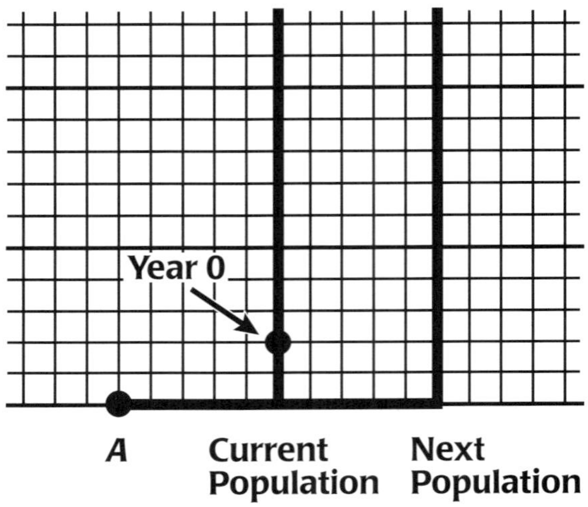

On **Student Activity Sheet 8,** there are two vertical lines: one is for the current caterpillar population and one is for next year's caterpillar population. The starting population (year 0) is marked along the current population line.

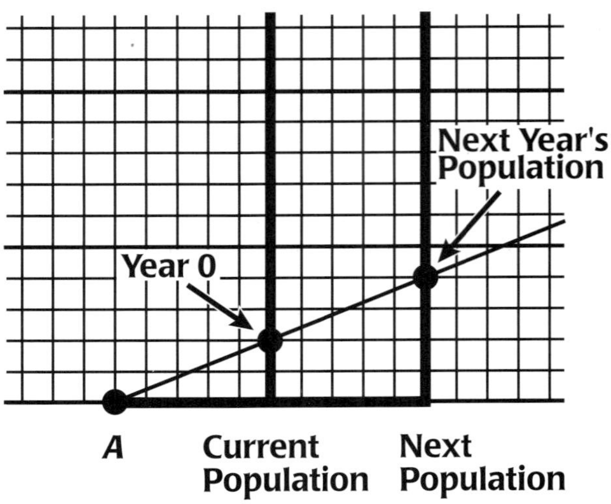

On **Student Activity Sheet 8,** draw a line through point *A* and the Current Population point, as shown on the left. The point at which your line intersects the Next Population line shows next year's population, the population for year 1.

3. What is the caterpillar population for year 1?

Solutions and Samples
of student work

3. The population for year 1 is four caterpillars.

Hints and Comments

Materials Student Activity Sheet 8 (one per student); transparency of Student Activity Sheet 8, optional (one per class); overhead projector, optional (one per class)

Overview The "explosive growth display" is introduced as a way to visualize exponential growth.

About the Mathematics The diagram that is introduced on this page and used on the next pages serves as a means of visualizing exponential growth. It is not a tool that students are expected to "master."

Planning You may need to spend some time introducing the explosive growth display. You may wish to use an overhead transparency of Student Activity Sheet 8 to model drawing the first line. Students may continue working in small groups on problem **3.**

Comments about the Problems

3. This is a matter of reading the information from the graph.

Mark and label the population for year 1 on the Current Population line, as shown on the right.

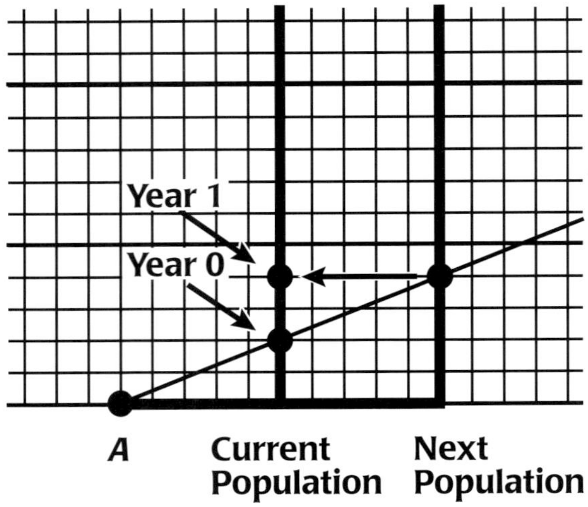

Draw a line through point *A* and the Current Population point for year 1, as shown on the right, to find the population for year 2.

4. a. What is the caterpillar population for year 2?

b. Mark and label the population for year 2 on the Current Population line.

c. Repeat this process to find the caterpillar populations for years 3 and 4.

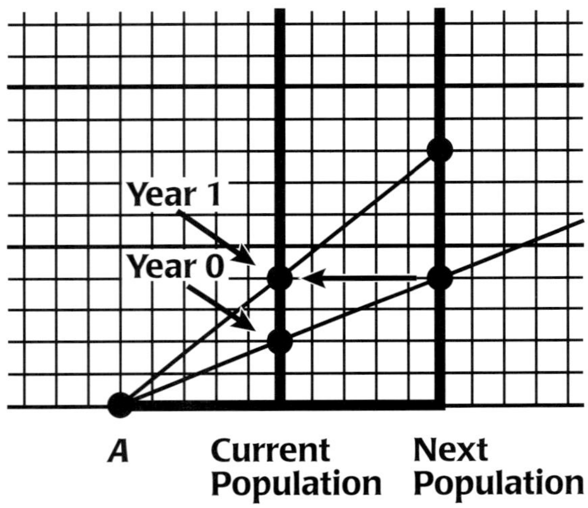

5. The population for year 5 won't fit on the grid provided on **Student Activity Sheet 8.** What is the population for year 5? How can you find this answer without extending the grid?

6. If you could extend the diagram on **Student Activity Sheet 8** with more diagrams that are the same size, how many would you need to plot a point for year 6?

Solutions and Samples
of student work

4. The population for year 2 is 8. The figure below shows the progression through year 4.

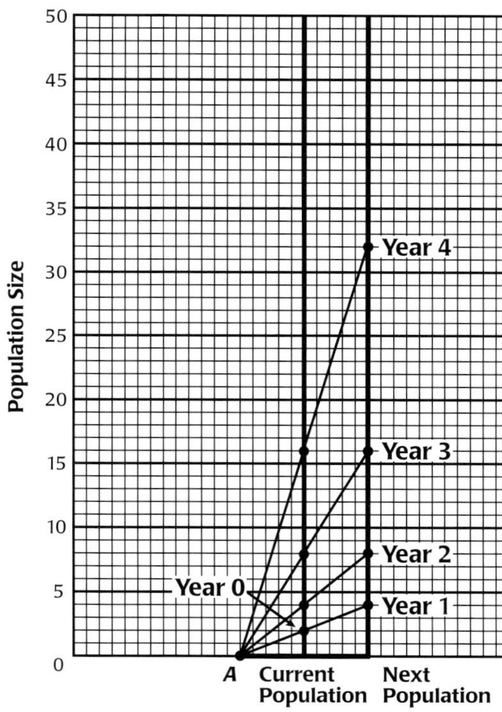

5. The population for year 5 is 64. You can find this by doubling the population of year 4.

6. The population for year 6 is 128, so you would need three diagrams representing 50 caterpillars each.

Hints and Comments

Materials Student Activity Sheet 8 (one per student)

Overview Students continue their exploration of the "explosive growth display."

Planning Students may continue working on problems **4–6** in small groups.

On **Student Activity Sheet 9,** point *A* is farther to the left than it is on **Student Activity Sheet 8.**

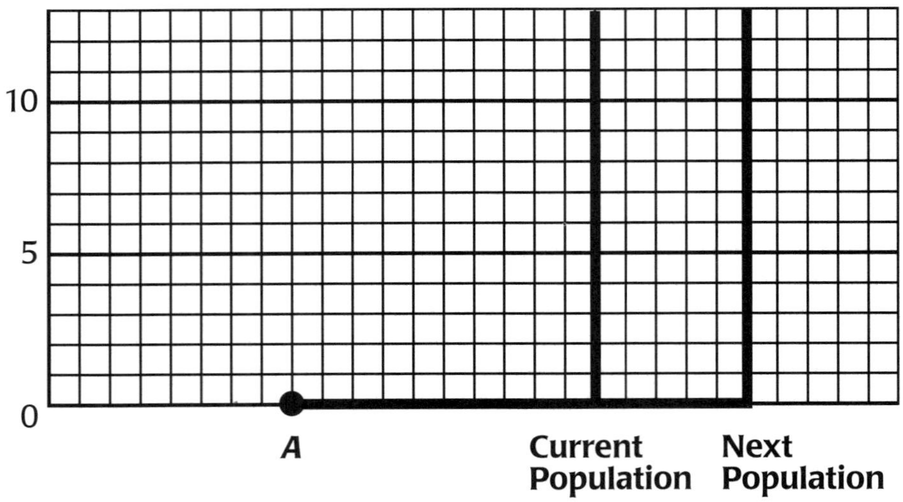

7. **a.** On **Student Activity Sheet 9,** make a point on the Current Population line for a starting population of size 2. (This is the same starting population that you worked with on **Student Activity Sheet 8.**) Find the caterpillar population for this new situation over several years.

   b. Is the growth rate for the diagram on **Student Activity Sheet 9** faster, slower, or the same as the growth rate shown on **Student Activity Sheet 8?** Explain.

A growth diagram is an easy way to visually calculate next year's population if you know this year's population and the growth factor. So far you have seen two growth diagrams with the same starting population but different growth factors.

8. Explain how you can adapt the growth diagram so that the growth rate is faster or slower without changing the starting population.

It is helpful to use a special kind of notation to represent the population size at a particular time. The population size at year 4 can be written as P(4), which is read as "P of 4," or it can be written as P_4, which is read as "P sub 4." The time interval does not always have to be years. It can be seconds, minutes, centuries—whatever is appropriate to the situation.

9. On another copy of **Student Activity Sheet 9,** find the caterpillar population over several years if P(0) = 8. Label each point on your growth diagram with this new notation for population size.

7. a.

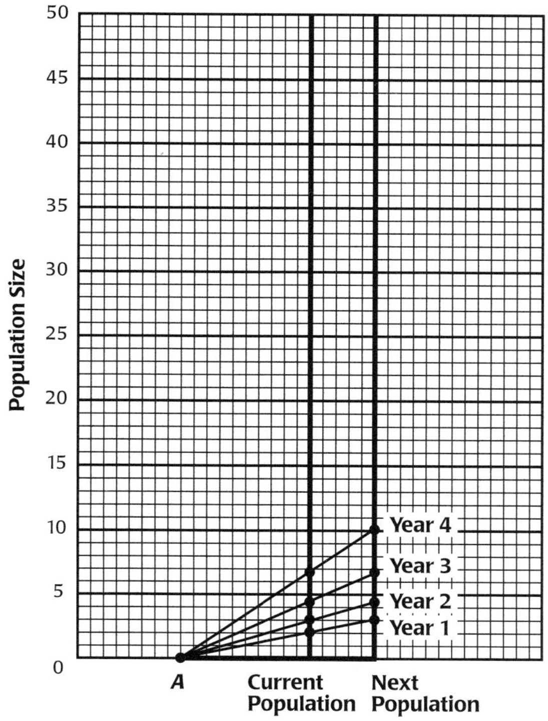

b. The growth rate for the diagram on Student Activity Sheet 9 is slower, because point *A* is farther from the Current Population line.

8. If you move point *A* closer to the Current Population line, the growth is faster. If you move point *A* away from the line, the growth is slower.

9.

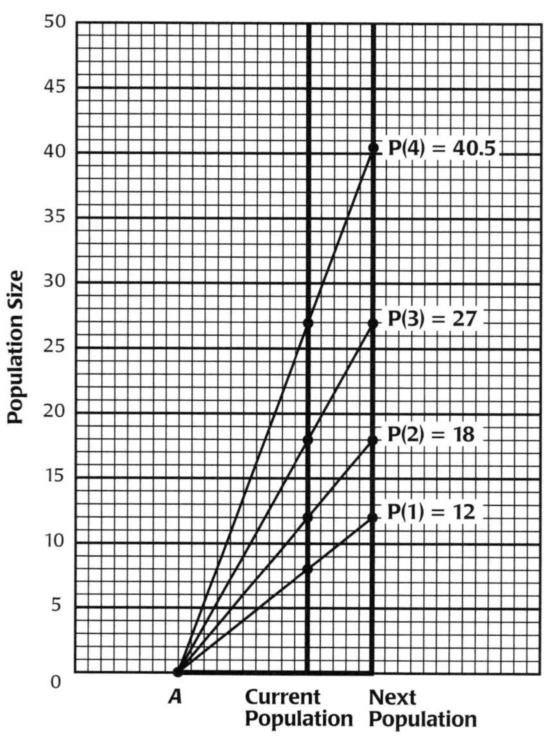

Materials Student Activity Sheets 8 and 9 (one of each per student)

Overview Students use the explosive growth display, but now for a different growth factor.

About the Mathematics The placement of point *A* on the horizontal axis determines the growth factor. When the distance between the Current Population line and the Next Population line is the same as the distance between point *A* and the Current Population line, the growth factor is 2. When *A* is farther from the Current Population line, the growth factor is smaller than 2, but it cannot be 1. When *A* is closer to the Current Population line, the growth factor is larger than 2. This can be explained by looking at the ratio between the distance from point *A* to the Current Population line and the distance from the Current Population line to the Next Population line. When the ratio is 1 (with equal distances), each population is double the previous population. When the ratio is 2:1, each population is 1.5 times the previous population. The growth factor is independent of the starting population.

Planning When you discuss problem **8,** use examples from this page and the previous pages to help students discover that the starting population does not influence the growth factor. The distance between the Current Population line and point *A* influences the growth factor. Students should have an intuitive understanding of this. Before students work on problem **9,** make sure that they understand the notations P(4) and P_4. Students may work on problems **7–9** individually. Problems **7** and **8** may be assigned as homework. Problem **9** can be used as an informal assessment.

Comments about the Problems

7–8. Homework These problems may be assigned as homework.

7. b. This problem draws students' attention to the relative *speeds* indicated by growth rates.

8. Students should develop an intuitive understanding of the effect of the distance from point *A* to the Current Population line.

9. Informal Assessment This problem assesses students' ability to interpret and create graphs representing exponential growth.

On the previous pages, you made three growth diagrams. For the first diagram, you had a graph and a table of the population sizes for the first five years. For the second and third diagrams, you knew only the starting populations. In the following problems, you will find the rate of growth for your third diagram by using arrow language.

10. Copy the following table into your notebook and fill in the value for each year's population starting with P(0) = 8 from your diagram from problem **9.**

Year	0	1	2	3
Population	8			

11. a. In your notebook, write the population values as an arrow string.

$$8 \xrightarrow{\times ?} \underline{\ ?\ } \xrightarrow{\times ?} \underline{\ ?\ } \xrightarrow{\times ?} \underline{\ ?\ }$$

b. What number should go above each arrow?

The number that you use above each arrow in the arrow string is the growth factor. The growth factor shows how fast the growth is occurring. For your arrow string, the same number should be above each arrow because the growth factor for a growth diagram is constant.

$$P(0) \xrightarrow{\times \text{ growth factor}} P(1) \xrightarrow{\times \text{ growth factor}} P(2) \xrightarrow{\times \text{ growth factor}} P(3)$$

With the growth diagram, each new population size is equal to the previous population size times the growth factor.

P(1) = growth factor × P(0),
P(2) = growth factor × P(1),
P(3) = growth factor × P(2), and so on.

12. Write a NEXT-CURRENT formula for your growth diagram from problem **9.**

Whenever each new value is found by applying the same formula to the preceding value, such as in a NEXT-CURRENT formula, the process is called *recursion* and the formula is called a recursive formula.

Solutions and Samples
of student work

10.

Year	0	1	2	3
Population	8	12	18	27

11. a. $8 \xrightarrow{\times 1.5} 12 \xrightarrow{\times 1.5} 18 \xrightarrow{\times 1.5} 27$

b. 1.5

12. NEXT POPULATION = CURRENT POPULATION × 1.5

Hints and Comments

Overview The term *growth factor* is introduced, and students explore ways to describe explosive growth with formulas.

About the Mathematics In the algebra units *Expressions and Formulas, Building Formulas,* and *Ups and Downs,* students worked with arrow language as a visual way to describe calculation rules. The arrow language used on this page is similar to a NEXT-CURRENT formula. The growth factor is the number by which the current amount is multiplied to get the next amount. NEXT-CURRENT formulas are also called recursive formulas. This term has been used in earlier algebra units.

Planning You may wish to have students work on problems **10–12** in small groups, and then discuss arrow language notation. Students should understand the relationship between explosive growth, arrow language, and the NEXT-CURRENT formula.

Comments about the Problems

12. A NEXT-CURRENT formula should be easy for students to find, since they have worked with this idea to find population values. The term *recursive formula* has been used in the units *Patterns and Symbols, Building Formulas, Ups and Downs, and Patterns and Figures.*

The idea of recursion may be easier to see if you make a card with two windows like the one shown on the right.

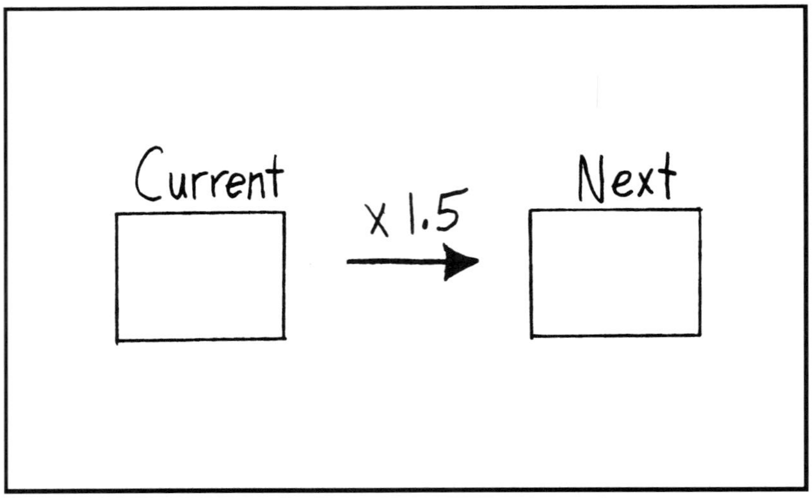

Sara uses her card to find the next two population values when the starting population is 8 and the growth factor is 1.5.

13. a. In your notebook, continue Sara's work for four more steps.

b. Explain how Sara's card works.

14. a. Suppose P(0) = 75 and the growth factor is 3.5. Find P(4) and explain how you found your answer.

b. Matt used his calculator to find P(14) for the growth described in part **a.** He then realized that he was supposed to find P(13), not P(14). What is a quick way for Matt to correct his mistake?

Solutions and Samples
of student work

13. a. $18 \xrightarrow{\times 1.5} 27 \xrightarrow{\times 1.5} 40.5 \xrightarrow{\times 1.5} 60.75 \xrightarrow{\times 1.5} 91.125$

b. Explanations will vary. Sample explanation:

The card is like a calculation rule; you put a number in the left box and by applying the rule (× 1.5), you get a number for the box on the right-hand side.

14. a. P(4) is 11,254.6875. Students should repeatedly multiply by 3.5.

P(0) = 75

P(1) = 75 × 3.5 = 262.5

P(2) = 262.5 × 3.5 = 918.75

P(3) = 918.75 × 3.5 = 3,215.625

P(4) = 3,215.625 × 3.5 = 11,254.6875

b. Matt can divide P(14) by 3.5 to get P(13) since P(14) = P(13) × 3.5.

Hints and Comments

Overview Students continue investigating explosive growth and recursive formulas.

About the Mathematics The card with two windows that is introduced on this page is a "one-arrow formula." It emphasizes the idea that the same operation is used repeatedly.

Planning After students have finished problems **13** and **14,** you can summarize this section using students' answers to these problems and the Summary text on the next page. Students may work on problems **13** and **14** individually. These problems may also be assigned as homework.

Comments about the Problems

13–14. Homework These problems may be assigned as homework.

13. Students should write a row of values starting with 18, and with each new value being the previous one times 1.5

14. Students may now discover (or rediscover) that using a NEXT-CURRENT formula can be time-consuming if you need to find a value that is higher in the sequence. This creates the need for a direct formula. In the next section, direct formulas will be introduced.

For any two consecutive numbers in an exponential sequence, the latter is the previous number multiplied by the growth factor. This implies that when the latter number is known, the previous number can be found by dividing the latter number by the growth factor.

Students should know that division is the reverse of multiplication.

Did You Know? Repeating a calculation can be done very easily on most calculators. By simply pushing on the = button, the last operation is repeated on the answer. To get a list of values for a NEXT-CURRENT formula, a calculator can thus be very helpful.

Summary

In this section, you examined explosive growth by working with graphs, growth diagrams, arrow language, and recursive formulas. With explosive growth, relatively slow initial growth soon gives way to increasingly rapid growth.

Explosive growth can be modeled by using arrow language, such as the following:

$$8 \xrightarrow{\times 1.5} 12 \xrightarrow{\times 1.5} 18 \xrightarrow{\times 1.5} 27$$

You can also illustrate explosive growth numerically as follows:

$P_0 = 8$
$P_1 = 8 \times 1.5$
$P_2 = 8 \times 1.5 \times 1.5$
•
•
•
$P_n = 8 \times 1.5 \times 1.5 \times 1.5 \times 1.5 \times 1.5 \ldots$ (*n* times)

To find each new population size, the previous population is multiplied by the growth factor. Because the same step is repeated each time to find a new value, explosive growth can be described by a recursive formula such as the following:

NEXT VALUE = CURRENT VALUE × 1.5

Summary Questions

15. Describe how you can determine whether growth is explosive or constant from each of the following:

a. a graph

b. a table

c. an equation

16. Give an example of a recursive formula and explain how it works.

Solutions and Samples
of student work

15. a. Graph:

Explosive growth—the graph is a curve that becomes steeper and steeper.

Constant growth—the graph is a straight line.

b. Table:

Explosive growth—each next number in the table is the previous number multiplied by the growth factor.

Constant growth—the differences between the numbers in the table are the same.

c. Equation:

Explosive growth—the equation has the form of $y = s(g^x)$, where s is the starting value and g is the growth factor.

Constant growth—the equation has the form of $y = ax + b$

16. Formulas will vary. Sample formulas:

Constant growth: NEXT = CURRENT + 3

Explosive growth: NEXT = CURRENT × 3

Hints and Comments

Overview Students read the Summary, which reviews models of explosive growth. They then complete two problems to show that they understand explosive and constant growth and recursive formulas.

About the Mathematics The mathematics addressed in this section includes the following: growth factors, explosive growth, arrow language, recursive formulas, graphs and tables, and the starting population. After this section, students should be able to describe explosive growth with tables, graphs, and recursive formulas.

Planning Have students read and discuss the Summary. They may work individually on problems **15** and **16.** These problems may be used for informal assessment.

After students complete Section D, you may assign appropriate activities from the Try This! section, located on pages 43–46 of the *Growth* Student Book, for homework.

Comments about the Problems

15. Informal Assessment This problem assesses students' ability to recognize and describe growth patterns presented in pictures, tables, and graphs; to interpret and create direct and recursive formulas for linear and exponential growth; to identify common traits within each type of formula: linear and exponential; and to understand the similarities and differences between linear and exponential growth.

16. Informal Assessment This problem assesses students' ability to interpret and create recursive formulas for linear and exponential growth.

Extension Ask students to think of an explosive growth situation. To present the situation, students should

- describe the growth factor,
- give the starting population,
- draw an explosive growth diagram and explain where point *A* is located, and
- write a recursive formula that describes the growth process.

Work Students Do

Students explore the ideas developed in Section D and extend them to formulas for exponential growth. They use growth factors to model the growth of real plants more accurately. They use graphing calculators to represent exponential formulas and investigate the effects of different growth factors and of different starting values on graphs.

Goals

Students will:

- interpret and create graphs and tables representing exponential growth processes;*
- interpret and create direct and recursive formulas for linear and exponential growth;
- understand the differences between recursive and direct formulas;
- identify common traits within each type of formula: linear, quadratic, cubic, and exponential;
- understand the similarities and differences between linear, quadratic, cubic, and exponential growth;*
- understand how changing a formula affects the graph and vice versa;*
- model a growth situation by choosing an appropriate formula and justify the choice;*
- recognize that a mathematical model usually does not fit a situation perfectly;
- recognize that growth processes can be simulated with a simplified mathematical model.

* *These goals are assessed in other sections of the unit.*

Pacing

- approximately three 45-minute class sessions

Vocabulary

- base
- exponent
- exponential equation
- exponential growth
- exponential notation
- power

About the Mathematics

An exponential function described with recursion takes the form of NEXT = CURRENT × constant. Described with a direct formula, it is $y = ac^x$, where c is a constant called the *growth factor, a* is the starting population, and x represents the time period. Exponential functions grow very quickly in the long run. When the growth factor is larger than 1, the growth is increasing, and when the growth factor is greater than 0 and less than 1, the growth is decreasing. The terms used with exponential notation (*power, base, and exponent*) are used in this section. At the end of this section, students should know what characterizes constant (or linear) growth and what characterizes exponential growth.

SECTION E. EXPONENTIAL GROWTH

Materials

- Student Activity Sheet 10, page 126 of the Teacher Guide (one per student)
- graphing calculators (or other graphing utility), pages 89, 91, 93, 95, and 97 of the Teacher Guide (one per student)
- graph paper, pages 89 and 97 of the Teacher Guide (several sheets per student)

Planning Instruction

You may wish to begin this section by reviewing what students did in Section A and discussing ways to model plant growth. Briefly revisit the differences between constant and explosive growth. This section is an extension of the previous one, so you may decide to move smoothly from Section D to this one without any formal introduction.

Students can work in small groups on problems 1–4 and 11–14. You may have them work on problems 5–10, 15, and 16 individually.

Problem 14 is optional. If time is a concern, it may be omitted or assigned as homework.

Homework

Problems 7 (page 88 of the Teacher Guide) and 8 (page 90 of the Teacher Guide) can be assigned as homework. After students complete Section E, you may assign appropriate activities from the Try This! section, located on pages 43–46 of the *Growth* Student Book. The Try This! activities reinforce the key mathematical concepts introduced in this section.

Planning Assessment

- Problem 8 can be used to informally assess students' ability to interpret and create direct and recursive formulas for exponential growth.
- Problem 15 can be used to informally assess students' ability to interpret and create direct and recursive formulas for linear and exponential growth; to understand the differences between recursive and direct formulas; and to identify common traits within each type of formula: linear and exponential.
- Problem 16 can be used to informally assess students' ability to create direct formulas for exponential growth. It also assesses their ability to recognize that a mathematical model usually does not fit a situation perfectly and to recognize that growth processes can be simulated with a simplified mathematical model.

E. EXPONENTIAL GROWTH

Simulating Plant Growth

You may have noticed a similarity between the growth of the Wisconsin fast plants that you observed in Section A and the graphs that you worked with in Section D that show explosive growth. In this section, you will develop an equation to represent explosive growth. To find an equation, first let's consider the growth of a simulated plant. Listed below are some assumptions that you need to make about the simulated plant.

- The plant is first measured when it is 32 millimeters tall, so P(0) = 32.
- The plant grows in a laboratory where environmental conditions are kept constant so that its growth does not fluctuate.
- The plant's height is measured at the end of each week.
- Based on other experiments with this plant, its growth can be modeled by the following formula:

NEXT WEEK'S HEIGHT = 1.5 × CURRENT WEEK'S HEIGHT

With the above information, you can graph the height of your simulated plant over time.

1. a. Use the information above to create a table of values to check against the graph shown on the right.

b. Suppose you measured the plant's height in the middle of the week instead of at the end of the week. Does the above NEXT-CURRENT formula still hold true? Explain.

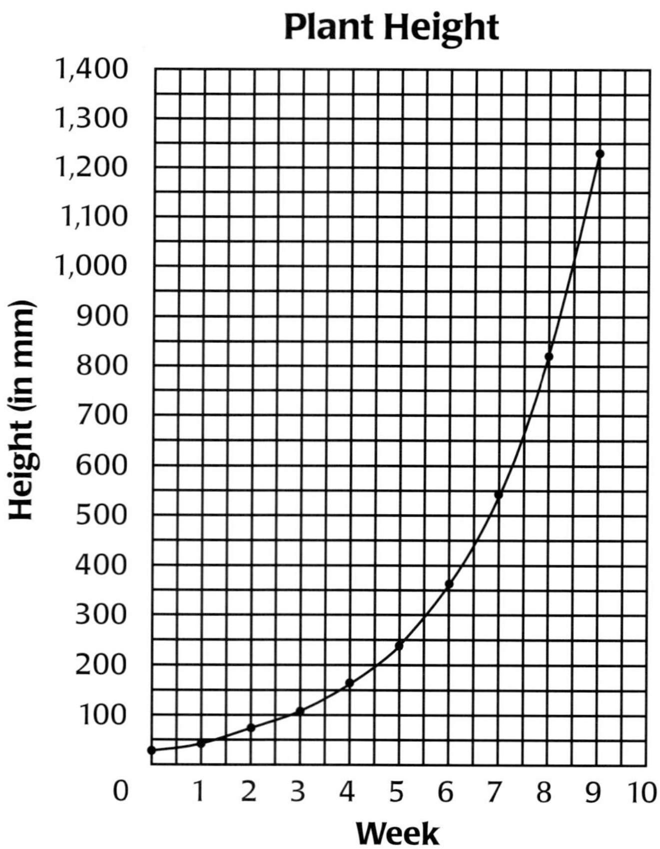

1. a.

Week	Plant Height (in mm)
Start	32
1	48
2	72
3	108
4	162
5	243
6	364.5
7	546.75
8	820.125
9	1,230.1875

The graphed heights seem to correspond with the heights obtained with the formula.

b.

Week	Plant Height (in mm)
.5	40
1.5	60
2.5	90
3.5	135
4.5	202.5
5.5	303.75
6.5	455.625
7.5	683.4375
8.5	1,025.15625

The graphed heights still seem to correspond with the heights obtained with the formula.

Overview The context of fast-growing plants from Section A is revisited, and students now use an exponential growth model.

About the Mathematics In this section, the term *exponential growth* is introduced to describe explosive growth as introduced in the previous section.

Planning Students can work in small groups on problems **1** and **2.** In these problems, students can use what they learned in the previous sections and apply it in the context of plant growth. You may want students to work through problems **1–4** before discussing their responses.

Comments about the Problems

1. Students should compare the values calculated with the formula to the values read from the graph. The values in the graph cannot be read very accurately, however.

b. The growth factor provides information about the growth during one week. That means that between any two moments that are a week apart, the plant grows 1.5 times larger.

Did You Know? Many processes in nature grow according to exponential models. A special growth factor appears often in nature; this is the irrational number *e*, which is equal to about 2.71. The number *e* also arises in situations in which interest is compounded continuously in bank accounts.

Exponential Equations

The growth processes of two different plants, plant A and plant B, are shown below.

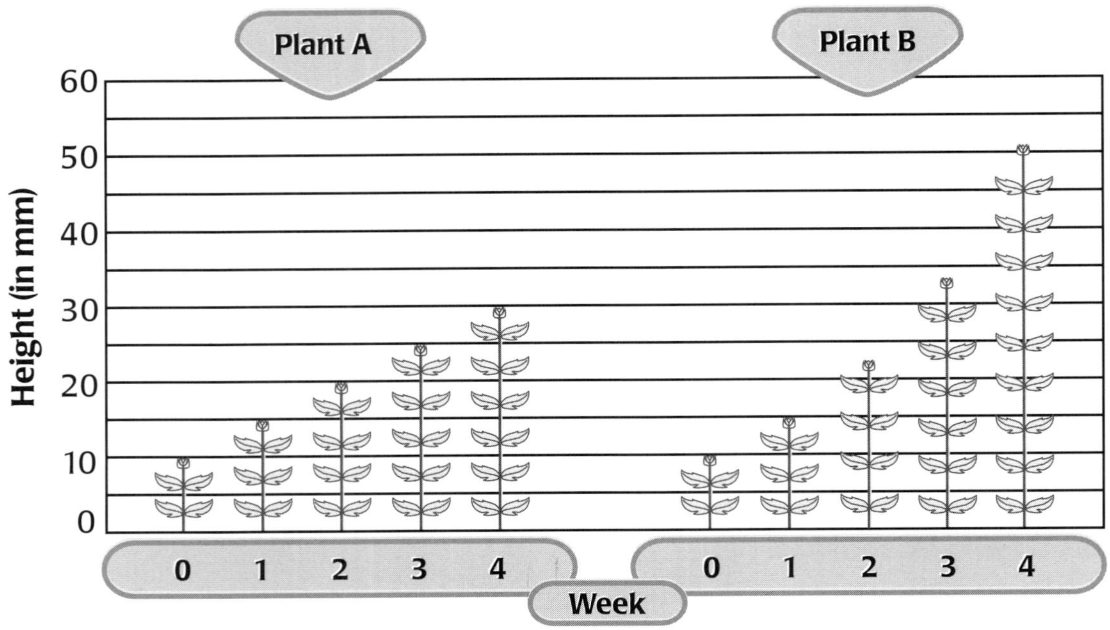

2. a. Describe the difference in the growth of these plants.

b. You can also describe the plants' growth with formulas. Copy and complete the following:

Plant A: P(0) = __?__
NEXT HEIGHT = CURRENT HEIGHT + __?__

Plant B: P(0) = __?__
NEXT HEIGHT = __?__ × CURRENT HEIGHT

In this unit, you have worked with two types of formulas: NEXT-CURRENT formulas and direct formulas.

With a NEXT-CURRENT formula, you must calculate the heights in successive steps. If you want to know the height for week 10, you must calculate the height, step-by step, for all of the weeks before week 10.

With a direct formula, you can find the height for week 10 with one calculation.

Solutions and Samples
of student work

2. a. Plant A shows linear growth—the increase is the same each week (5 mm).

Plant B shows explosive growth—each new height is that of the previous week multiplied by 1.5.

b. Plant A: P(0) = 10 mm

NEXT HEIGHT = CURRENT HEIGHT + 5

Plant B: P(0) = 10 mm

NEXT HEIGHT = CURRENT HEIGHT × 1.5

Hints and Comments

Overview Students compare the growth of two plants represented by pictures and write recursive formulas to describe the growth processes.

About the Mathematics Constant (or linear) growth is characterized by a constant amount that is added during each time period. In a recursive formula, this shows as NEXT = CURRENT + constant. Explosive (or exponential) growth is characterized by a constant number by which the current value is multiplied to get the next value. In a recursive formula, this shows as NEXT = CURRENT × constant. Linear growth is additive; exponential growth is multiplicative.

Planning Have students continue working in small groups on problem **2.** In previous sections and in previous units (*Ups and Downs, Building Formulas,* and *Graphing Equations*), students worked with recursive and direct formulas for constant growth.

Comments about the Problems

2. Encourage students to describe the growth in their own words and also to describe the growth using terminology from the unit. When you discuss this problem (which is best discussed after students complete problem **4**), talk about the differences between the two formulas: constant growth is adding a constant, and explosive growth is multiplying by a constant.

If the amount of time is measured in weeks, a direct formula for plant A is the following:

height = 10 + 5 × time

or

$h = 10 + 5t$

3. Check this direct formula for plant A by creating a table of values and comparing it to the graph on page 36.

You can record the data for plant B in a table.

Time (in weeks)	Calculation for the Height	Height (in mm)
0	10	10
1	10 x 1.5	15
2	10 x 1.5 x 1.5	22.5
3	10 x 1.5 x 1.5 x 1.5	33.75
4	10 x 1.5 x 1.5 x 1.5 x 1.5	50.625

4. How would you calculate the height for plant B at 10 weeks?

The calculation for problem **4** is easier to write if you use exponential notation. With exponential notation, 7 × 7 × 7 × 7 × 7 can be written as 7^5; 7 is the base of 7^5, and 5 is the exponent of 7^5.

You can use this notation to make a new table for plant B.

Time (in weeks)	Calculation for the Height	Height (in mm)
0	10	10
1	10 x 1.5	15
2	10 x $(1.5)^2$	22.5
3	10 x $(1.5)^3$	33.75
4	10 x $(1.5)^4$	50.625

5. a. What is the calculation for the height (h) when time (t) = 10?

b. What is the calculation for h when t = 100?

c. Write a direct formula that allows you to calculate h for any value of t.

3.

Time (in weeks)	Height from Direct Formula	Height from Graph
0	10 mm	10 mm
1	15 mm	15 mm
2	20 mm	20 mm
3	25 mm	25 mm
4	30 mm	30 mm

4. $10 \times 1.5 \times 1.5 \times 1.5 \times 1.5 \times 1.5 \times 1.5 \times 1.5 \times 1.5 \times 1.5 \times 1.5$, or $10 \times (1.5)^{10}$

5. a. $h = 10 \text{ mm} \times (1.5)^{10} \approx 576.65 \text{ mm}$

b. $h = 10 \times (1.5)^{100} \approx 4.066 \times 10^{18}$ mm (This is more than 4,000 million km, or more than 4 billion km.)

c. $h = 10 \times (1.5)^t$

The h is the height in millimeters, and the t is the time in weeks.

Overview Students investigate direct formulas for linear growth and start thinking about direct formulas for exponential growth. They revisit exponential notation.

About the Mathematics Students should know the terminology used in exponential notation as defined on this page.

Planning Have students continue working in small groups on problems **3** and **4.** After problem **4,** you may have a class discussion about the different kinds of growth and about recursive and direct formulas. Then spend some time on the terminology that is involved with exponential notation. Students may remember the terms from the unit *Powers of Ten.* Have students work individually on problem **5.** A direct formula for exponential growth is introduced on the next page.

Comments about the Problems

3. Have students explain in their own words the meanings of 10 and 5 in this formula.

4. Some students may start thinking about finding a direct formula instead of a recursive formula in this problem. Do not push students toward a direct formula yet, however.

5. Problem **5** is critical; students must be able to use exponential notation with understanding in the context of growth. You may need to spend some extra time in class on exponential notation.

Although the answers to the calculations are given in the solutions column, students need not compute these values. It is sufficient if they can write the expressions that represent the calculations.

Extension As an extension to problem **5,** you may discuss the answers to the calculations and talk about how quickly the numbers grow. If your students need some extra practice working with exponential notation, you may want to write some powers on the chalkboard and have students indicate what the base is and what the exponent is for each. You may also give them some bases and exponents and ask students to write the powers that go with these parts.

Even though it is very easy to write the calculations for plant B with exponential notation, you still need an easy way to perform the calculations. There is a special key on your calculator for computing a power. Since calculators differ, you may want to investigate how to use the power key on your calculator by verifying the following calculations:

$2^4 = 16$ $3^3 = 27$ 1.6^3 or $(1.6)^3 = 4.096$

6. a. In your notebook, extend the table shown at the bottom of page 37 to include $t = 5$ and $t = 6$.

b. Compare your formula from problem **5c** to the NEXT-CURRENT formula from problem **2b.**

In equations like those shown below, the variable t is an exponent, so the equations are called *exponential equations.* The growth that they describe is called *exponential growth.*

$h = 10(1.5)^t$ $h = 8.5(3)^t$ $h = 2^t$

Martha grew two different kinds of plants in her science class. She labeled them plant C and plant D. For plant C, she found that the height at a given time could be described by the formula $h = 20 \times (1.2)^t$, where t is any number between 0 and 11 ($0 < t < 11$). Martha measured t in weeks and h in millimeters.

7. a. On graph paper, choose a suitable scale and make a graph for plant C that shows its height over time.

b. For plant D, the formula is $h = 20 + 4t$. Draw the graph for plant D on the same diagram as plant C.

c. What kind of growth is displayed by plant C? by plant D?

d. When does plant C become taller than plant D?

e. How will your answer to part **c** change if you make the slope of the graph for plant D steeper?

6. a. For $t = 5$, the height is about 75.94 mm, and for $t = 6$, the height is about 113.9 mm.

b. Comparisons will vary. Sample response:

The recursive formula NEXT = CURRENT × 1.5 is related to the direct formula $h = 10 \times (1.5)^t$ in that the starting height is 10 millimeters and the recursive formula is applied t times.

7. a–b.

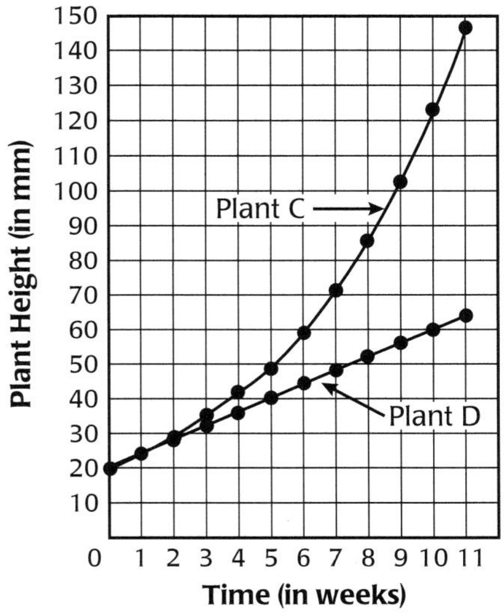

c. Plant C: exponential growth

Plant D: linear growth

d. According to the graph, plant C becomes taller than plant D at some time between one and two weeks.

e. Answers will vary, but students should note that something with exponential growth will always outgrow something with linear growth, assuming enough time passes. If plant D had a steeper slope, it would take longer for plant C to overtake it.

Materials graph paper (at least one sheet per student), scientific calculators, optional (one per student)

Overview Students investigate recursive and direct formulas for exponential growth.

About the Mathematics The general formula for constant growth is $y = sx + i$, where s is the slope of the line (or the constant that is added) and i is the y-intercept. Another often-used formula is $y = mx + b$. The general formal for exponential growth is $y = s(g)^x$, where s is the starting value and g is the growth factor. In the context of the problem situations in this section, y represents the height (h), and x represents the time (t). In problem **7**, a graphical way of solving equations is applied; students find the point of intersection by looking at the graph instead of solving systems of equations.

Planning You may want to discuss how to use the calculator to calculate powers and do the three calculations using the demonstration calculator. Then you can have students work individually on problem **6;** following this, talk about the term *exponential growth* and discuss the examples presented between problems **6** and **7.** Students may work individually on problem **7,** which may be assigned as homework.

Comments about the Problems

6. b. Students should recognize that the growth factor is in both formulas, but that the NEXT-CURRENT formula does not give any information about the starting value.

7. Homework This problem may be assigned as homework. After students have completed the problem, talk about the differences between the two graphs (the line and the curve). Ask students how to find the point of intersection and what that point means in the given context. Discuss why, in the long run, exponential growth always wins over linear growth, no matter how steep the slope of the line.

Suppose the height of plant E at time *t* is described by the formula $h = 20 \times (1.4)^t$.

8. **a.** What does the number 20 tell you about the plant?

b. What does the number 1.4 tell you about the plant?

c. What is a possible NEXT-CURRENT formula for this plant?

The number 1.4 in the direct and NEXT-CURRENT formulas is called the growth factor. In the following problem, you will explore how the growth factor affects the graph of an exponential equation.

9. **a.** On your graphing calculator, graph each of the following equations: $Y = 2^X$, $Y = 5^X$, and $Y = 10^X$. Use the table feature of your calculator to determine an appropriate viewing window.

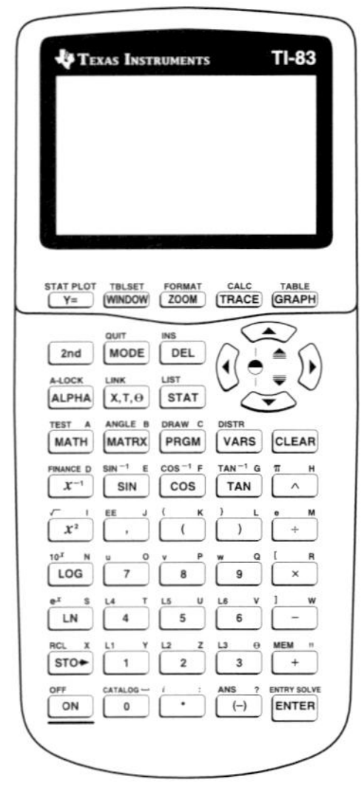

b. How would you expect the graph of $Y = 7^X$ to compare to the graphs from part **a?** What about $Y = (0.5)^X$ and $Y = (0.75)^X$? Check your answers with your graphing calculator. (*Hint:* You might need to change your viewing window to see all of these graphs in the same window.)

c. Use your calculator to experiment with some other growth factors and write a paragraph about what you discover.

Now that you have investigated the effect of changing the growth factor, let's investigate how the graph changes if the starting point changes.

10. Consider three plants, each with a growth factor of 2, but one with a starting height of 5, one with a starting height of 10, and one with a starting height of 15.

a. Write a direct formula to describe each plant.

b. Use your graphing calculator to investigate the graph for each of these plants.

c. Write a paragraph describing your discoveries about how the starting point affects the graph of an exponential equation.

Solutions and Samples
of student work

8. a. The plant is 20 units tall when measurements begin.

b. The growth factor is 1.4. This is the number by which the height is multiplied for each period of time.

c. NEXT HEIGHT = CURRENT HEIGHT × 1.4

9. a.

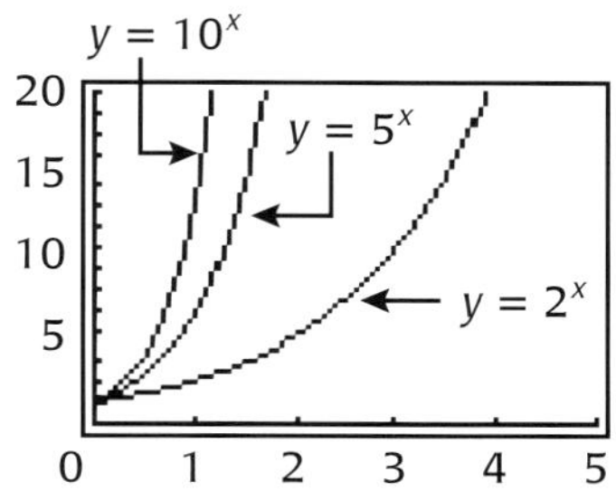

b. The graph of $y = 7^x$ would be between the graphs of $y = 5^x$ and $y = 10^x$. The graph of $y = 0.5^x$ would be under the other graphs. The graph of $y = 0.75^x$ would be between $y = 0.5^x$ and $y = 2^x$. All graphs contain (0, 1).

c. Answers will vary, depending on the growth factors that students use. Students should note the positions of the graphs relative to each other or to the graphs from parts **a** and **b.** When the growth factor is between 0 and 1, the growth is decreasing.

10. a. $y = 5(2^x)$

$y = 10(2^x)$

$y = 15(2^x)$

b.

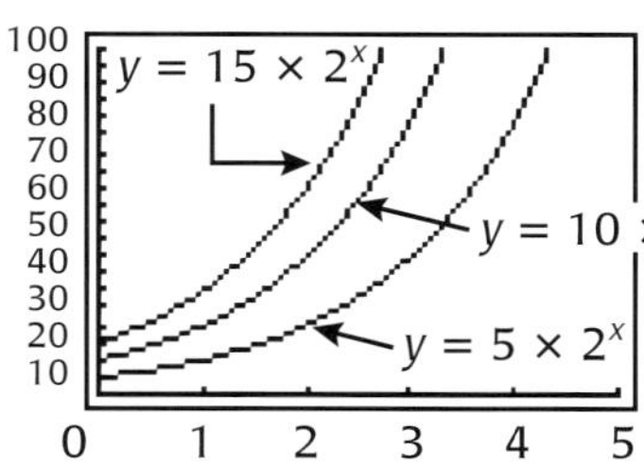

c. Descriptions will vary. Sample description:

Each of the three formulas describes doubling, but since the starting values are different, the graphs also differ, but each graph increases (doubles) rapidly.

Hints and Comments

Materials graphing calculators (one per student)

Overview Students investigate direct formulas for exponential growth and make the connection with the growth factor. They investigate the effects of different growth factors and of different starting values.

About the Mathematics The general formula for exponential growth is $y = s(g)^x$. When the growth factor g is the same, the starting value s influences where the graph intercepts the y-axis. When the growth factor (g) is larger than 1, the graph shows explosive growth; faster and faster, or in other words, increasing growth. When $0 < g < 1$, the graph shows decreasing growth; the growth goes slower and slower.

Planning Students may work individually on problems **8–10.** You may assign problem **8** as homework. You may also use problem **8** as an informal assessment. If you do not have graphing calculators available, you can draw graphs for problems **9** and **10** on the chalkboard as a demonstration. Discuss the different graphs with students. They should "play" with graphs to discover the effect of changing the starting value or growth factor.

Comments about the Problems

8. Informal Assessment This problem assesses students' ability to interpret and create direct and recursive formulas for exponential growth. It can also be assigned as homework.

9. In this problem, students investigate the effect of changing the growth factor.

10. Here, students investigate the effect of changing the starting value.

Shown below is a graph of plant E. The formula for this plant is $h = 20(1.4)^t$.

11. a. Use the above formula to calculate the value of h when $t = 1, 2, 3, 4, 5,$ and 6. Record your answers in the table on **Student Activity Sheet 10.**

b. If you were shown the graph without the formula, how could you find the growth factor?

Plant E: $h = 20(1.4)^t$

Time (in weeks)	0	1	2	3	4	5	6
Height (in mm)	20						

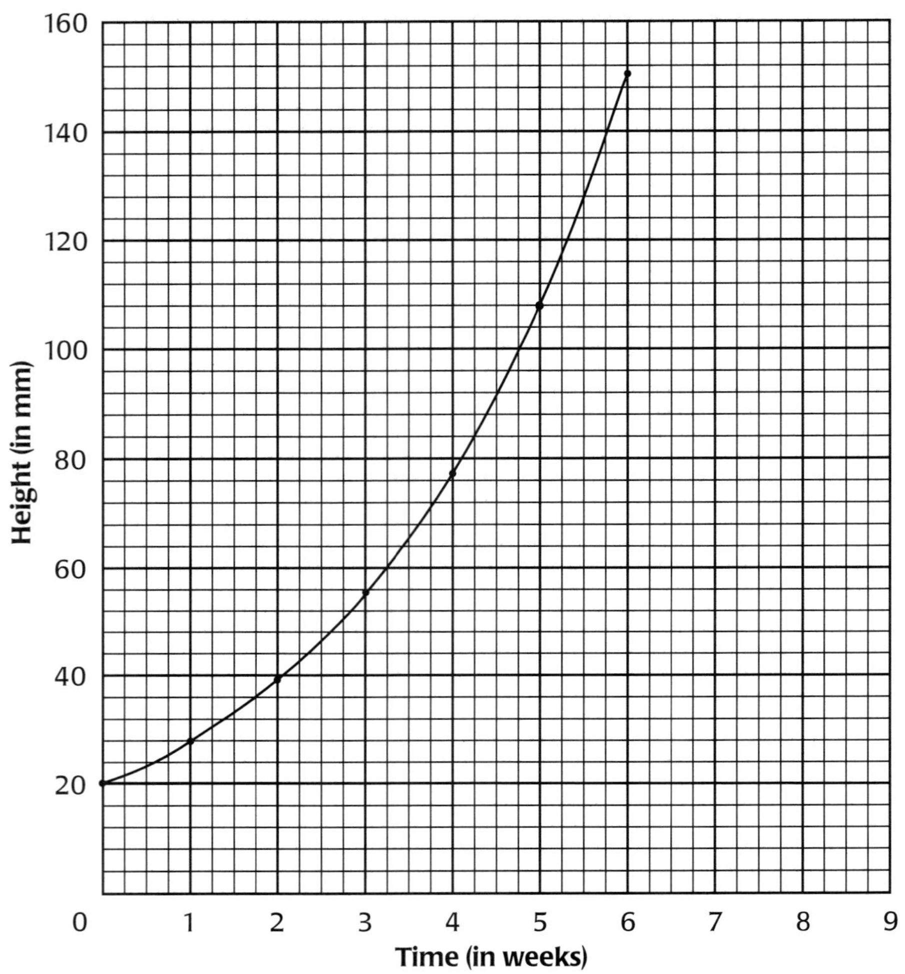

Now use your graphing calculator to investigate what happens to the graph of $h = 20(1.4)^t$ as t changes.

12. Graph $h = 20(1.4)^t$ on your graphing calculator and choose an appropriate viewing window.

a. If $t = 0$, what is h? Where is this point on the graph? What does this tell you about the value of $(1.4)^0$?

b. If $t = 2.5$, what is h? Where is this point on the graph? Explain why your calculated value of h seems reasonable.

c. For plant E, $t = 0$ represents the starting time for the observations. The plant has reached a height of 20 millimeters by the starting time. Calculate h for $t = -1$ and explain what this could represent.

Solutions and Samples
of student work

11. a. **Plant E: $h = 20(1.4)^t$**

Time (in weeks)	0	1	2	3	4	5	6
Height (in mm)	20	28	39.2	54.9	76.8	107.6	150.6

b. Answers will vary. Sample response:

The value at $t = 0$ is known (20 mm). You can read the value at $t = 1$ from the graph and divide this value by 20 to determine the growth factor.

12. a. When $t = 0$, $h = 20$, so $(1.4)^0 = 1$.

b. When $t = 2.5$, $h = 20(1.4)^{2.5}$

$h \approx 46.4$ mm

Explanations may vary. Students may note that the point on the graph for $t = 2.5$ has an h value of about 46.

c. For $t = -1$,

$h = 20(1.4)^{-1} = 20\left(\frac{1}{1.4}\right)$

$h = 20 \times 0.714$

$h \approx 14.29$ mm

This represents the height one week before you start measuring.

Hints and Comments

Materials Student Activity Sheet 10 (one per student); graphing calculators (one per student)

Overview Students investigate the formula $h = 20(1.4)^t$.

About the Mathematics In the formula $h = 20(1.4)^t$, 20 is the starting value; it is the height at $t = 0$. The number 1.4 is the growth factor, meaning that during each time period (weeks in this context), the height is multiplied by 1.4. In the context of growing plants, it is meaningful to look at the height at any moment of time. It is a continuous process.

It is possible to determine the growth factor by looking at the values in a table or a graph. You can look at two consecutive time periods and compare the heights. By dividing the height at the latter time by the height at the former time, you get the growth factor.

Planning You can have students work on problems **11–12** in small groups.

Comments about the Problems

11. b. When you discuss this problem, you may want to refer to problem **14** in Section D (page 33 of the Student Book). It is important that students understand that the height at the NEXT time period is the height of the CURRENT moment multiplied by the growth factor. So, there are three values (current height, next height, and growth factor), and when two values are known, the third can be found.

12. a. By convention, any value to the power 0 is equal to 1. Students may remember this from the unit *Powers of Ten.*

c. A negative exponent of -1 in this context means one week before $t = 0$.

Extension You may want to talk with students about the mathematical meaning of negative exponents as it is introduced in *Powers of Ten* (a negative exponent means 1 divided by the power). This, however, is not essential for completing this unit.

Suppose you grow another plant, plant F. The growth conditions for this plant are different and the height of plant F at any time is only half of the height of plant E at the same time.

13. a. On the graph on **Student Activity Sheet 10,** graph plant F.

b. Find a direct formula for the height of plant F.

c. Study the graphs on **Student Activity Sheet 10** for plants E and F. Which of the following statements do you think is correct? Explain.

i. Plant F is growing more slowly than plant E because the height of plant F increases less than the height of plant E between $t = 0$ and $t = 6$.

ii. Plant F is not growing more slowly than plant E because the growth factor is the same for both plants.

14. Plant G has a height of 20 millimeters at $t = 0$ and a height of 75 millimeters at $t = 6$. Based on these results, the growth factor for plant G must be different than the growth factor for plants E and F. Use your calculator and the guess-and-check method to determine an approximate growth factor for plant G.

Solutions and Samples
of student work

13. a.

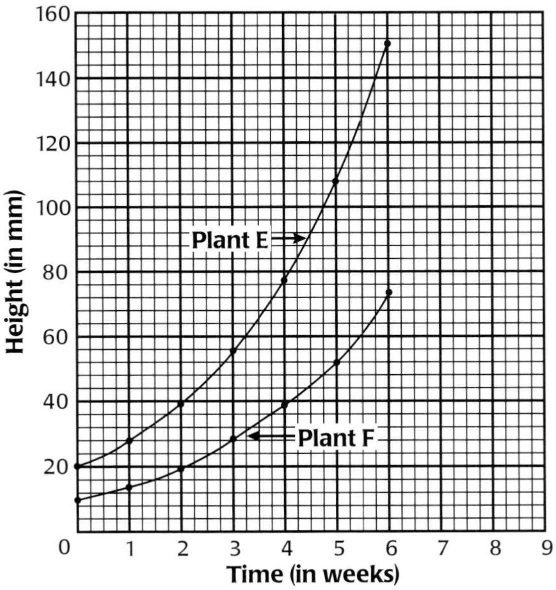

b. Answers may vary. Sample formulas:

- $h = \frac{1}{2} \times 20(1.4)^t$
- $h = 10(1.4)^t$

c. Statement **i** is correct when you look at the height only. Statement **ii** is correct when you look only at the growth factors.

14. The growth factor for plant G is about 1.25. Students should realize that $20 \times g \times g \times g \times g \times g \times g = 75$, where g represents the growth factor.

$20 \times g^6 = 75$

$g^6 = \frac{75}{20} = 3.75$

So, g must be between 1 and 2; the guess-and-check method should lead to a value of about 1.25.

Hints and Comments

Materials Student Activity Sheet 10 (one per student); graphing calculators (one per student)

Overview Students compare the growth of three plants.

About the Mathematics When the growth of one plant is half (or any other factor) that of a second plant, this does not mean that the growth factor is half. It means that the total growth of one plant is half that of the other.

Planning When you discuss problems **13** and **14,** talk about the effects of changing graphs by changing the formula, and vice versa. You may want students to work individually on problems **13** and **14.** Problem **14** is optional. If time is a concern, you may omit it or assign it as an extra challenge homework problem.

Comments about the Problems

13. Some students may think that the growth factor is halved. If students come up with this idea, you may use an example to illustrate that this is not the case. When the growth factor is 0.7, the graph looks completely different than when the growth factor is 1.4.

Students should realize that both statements can be true; it depends on what you look at. Students' justifications for their answers are more important than the answers themselves.

14. This problem is optional. You may use it for extra credit. Students should realize that between $t = 0$ and $t = 6$, there are 6 steps, thus six multiplications by the growth factor; in exponential notation, this is (growth factor)^6.

Summary

Explosive growth can be described by a NEXT-CURRENT formula of the following form:

$$\text{NEXT HEIGHT} = \text{CURRENT HEIGHT} \times \text{growth factor}$$

Explosive growth can also be called exponential growth because the direct formula for explosive growth is an exponential equation. The form for this exponential equation is as follows:

$$\text{height} = \text{starting height} \times (\text{growth factor})^{\text{time}}$$

The growth factor describes how quickly the height of the plant increases—the larger the growth factor, the more explosive the growth.

Summary Questions

15. Compare the NEXT-CURRENT formula for exponential growth with the direct formula for exponential growth.

So far in this section, all of the plants that you have worked with have been imaginary. Now let's return to the Wisconsin fast plants that you studied in Section A.

16. **a.** On graph paper, plot the median points for plants A–L from page 6 and draw a curve through the points. (You did this for problem **15b** on page 7.)

b. Using your graph and the guess-and-check method, determine an approximate growth factor for the curve that you drew.

c. Using your graphing calculator, check your answer to part **b** and determine a value for the growth factor to the nearest hundredth.

d. Write a direct formula for the height of a Wisconsin fast plant.

15. Comparisons will vary. Sample response:

The NEXT-CURRENT formula (the recursive formula) shows what happens in each step of the growth process: the multiplication of the growth factor.

The direct formula shows directly by how many times the growth factor is to be multiplied. It also shows the starting height.

16. a.

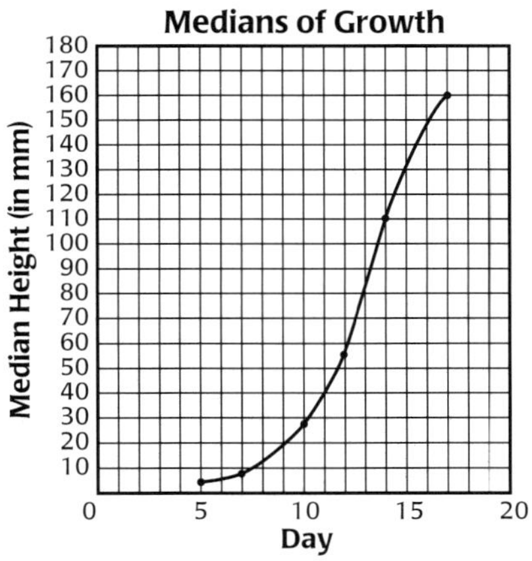

b. Answers will vary. Students can use any interval of one day to check the growth factor; for example, between days 10 and 11, the factor appears to be around 1.4:

h (day 10) = 27.5 mm

h (day 11) ≈ 38 mm

$27.5 \times g \approx 38$

$g \approx \frac{38}{27.5}$

$g \approx 1.4$

c. The growth factor to the nearest hundredth is 1.38. Students should plot the median values and use exponential regression to find the best-fitting graph.

d. $h = 4(1.38)^t$

You start measuring from day 5, when the height is about 4 mm, so on day 5, $t = 0$; on day 6, $t = 1$; and so on.

Materials graphing calculators (one per student); graph paper (at least one sheet per student)

Overview Students read and discuss the Summary, which reviews what students have learned about exponential growth. They then find a formula that represents the plant growth they studied in Section A.

About the Mathematics By the end of this section, students should have developed a conceptual understanding of exponential growth, its formulas and graphs, and how it relates to linear growth. The mathematics that students have dealt with in this section includes the following:

- NEXT-CURRENT and direct formulas for linear and exponential growth;
- the effects of different starting values and different growth factors; and
- finding a formula to describe exponential growth.

Planning Problems **15** and **16** can be used for informal assessment. After students complete Section E, you may assign appropriate activities from the Try This! section, located on pages 43–46 of the *Growth* Student Book, for homework.

Comments about the Problems

15. Informal Assessment This problem assesses students' ability to interpret and create direct and recursive formulas for linear and exponential growth; to understand the differences between recursive and direct formulas; and to identify common traits within each type of formula: linear and exponential.

16. Informal Assessment This problem assesses students' ability to create direct formulas for exponential growth. It also assesses their ability to recognize that a mathematical model usually does not fit a situation perfectly and to recognize that growth processes can be simulated with a simplified mathematical model.

c. If no graphing calculators are available, students may use graph paper. They may be inclined to fill in the growth factor they found in problem **15b** and check the result visually. Encourage them to check the growth factor by entering the median points from the graph in the graphing calculator.

Assessment Overview

Students work on three assessments activities that you can use to collect additional information about what each student knows about linear, exponential, and other growth processes and modeling and simulating growth processes.

Goals	Assessment Opportunities
• recognize and describe growth patterns presented in pictures, tables, and graphs	Money, Money, Money Strange Creatures A Growth Experiment
• interpret and create graphs and tables representing linear, quadratic, cubic, and exponential growth processes	Money, Money, Money A Growth Experiment Strange Creatures
• interpret and create direct and recursive formulas for linear and exponential growth	Money, Money, Money Strange Creatures A Growth Experiment
• identify the common traits within each type of formula: linear, quadratic, cubic, and exponential	Strange Creatures A Growth Experiment
• understand the similarities and differences between linear, quadratic, cubic, and exponential growth	Strange Creatures A Growth Experiment
• model a growth situation by choosing an appropriate formula and justify the choice	Money, Money, Money Strange Creatures A Growth Experiment
• recognize that a mathematical model usually does not fit a situation perfectly	A Growth Experiment
• recognize that growth processes can be simulated with a simplified mathematical model	Money, Money, Money Strange Creatures A Growth Experiment

Pacing

When combined, the three assessment activities will take approximately two or three 45-minute class sessions. See the Planning Assessment section for further suggestions as to how you might use the assessment activities.

About the Mathematics

The three assessment activities evaluate the major goals of the *Growth* unit. The final questions in each assessment also assess higher-order goals such as generalizing, reflecting, and reasoning. Refer to the Goals and Assessment Opportunities sections on the previous page for information regarding the specific goals assessed in each assessment activity. Students may use different strategies to solve each problem. Their choice of strategy may indicate their level of comprehension of the problem. Consider how well students' strategies address the problem, as well as how successful students are at applying their strategies to the problem-solving process.

Materials

- Assessments, pages 128–134 of the Teacher Guide (one of each per student)
- calculators, pages 101, 103, and 107 of the Teacher Guide, optional (one per student)
- graphing calculators, pages 111 and 113 of the Teacher Guide, optional (one per student)
- rulers, pages 111 and 113 of the Teacher Guide (one per student)
- colored pencils, pages 111 and 113 of the Teacher Guide (three different colors per student)

Planning Assessment

You may want students to work on these assessments individually so that you can evaluate each student's understanding and abilities. Make sure that you allow enough time for students to complete the assessment activities. Students are free to solve each problem in their own way. They may choose to use any of the models introduced and developed in this unit to solve problems that do not call for a specific model.

The focus of the three assessment activities is on exponential growth, although informal understanding of other kinds of growth processes is also assessed. You may choose to combine any two of the three assessment activities as one end-of-unit assessment. The Strange Creatures assessment may be used anytime after students finish Section C.

Scoring

Answers are provided for all assessment problems. The method of scoring the problems depends on the types of questions in each assessment. Most questions require students to explain their reasoning or justify their answers. For these questions, the reasoning used by the students in solving the problems as well as the correctness of the answers should be considered as part of your grading scheme. A holistic scoring plan can be used to evaluate an entire task. For example, after reviewing a student's work, you may assign a key word such as *emerging, developing, accomplishing,* or *exceeding* to describe the student's mathematical problem-solving, reasoning, and communication.

On other tasks, it may be more appropriate to assign point values for each response. Student progress toward the goals of the unit should also be considered. Descriptive statements that include details of a student's solution to an assessment activity can be recorded. These statements could provide insight into a student's progress toward a specific goal of the unit. Descriptive statements are often more informative than recording only a score and can be used to document student growth in mathematics over time.

MONEY, MONEY, MONEY

Use additional paper as needed.

Jim often daydreams about money. In his last daydream, he deposited a certain amount of money in his bank account and left it there for a two-year period. Here is a picture of Jim's daydream.

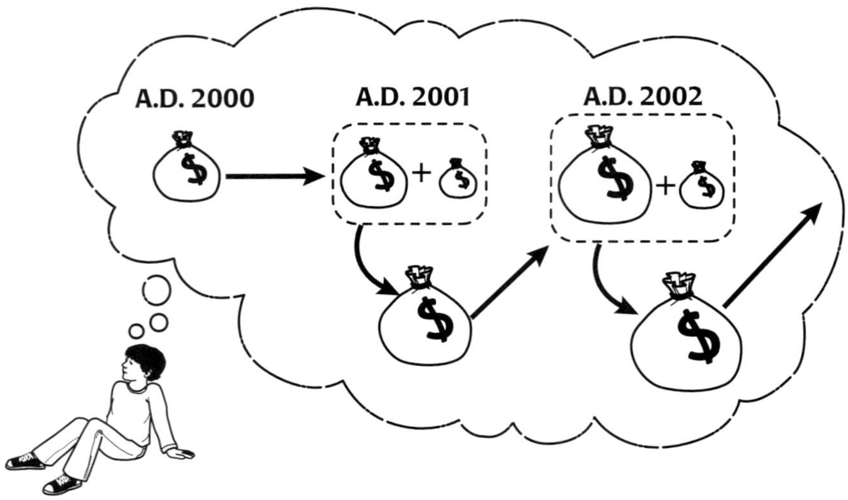

1. Describe what happened in this daydream.

Jim's original bank account deposit in the year A.D. 2000 was $1,000. His bank offers a 6% interest rate compounded annually. This means that when the money has been in the account for a year, Jim gets 6% interest on the sum of the original amount. At the end of the second year, he earns 6% interest on the amount he had at the beginning of that year.

2. Write the correct amount of money on each bag shown in the picture below.

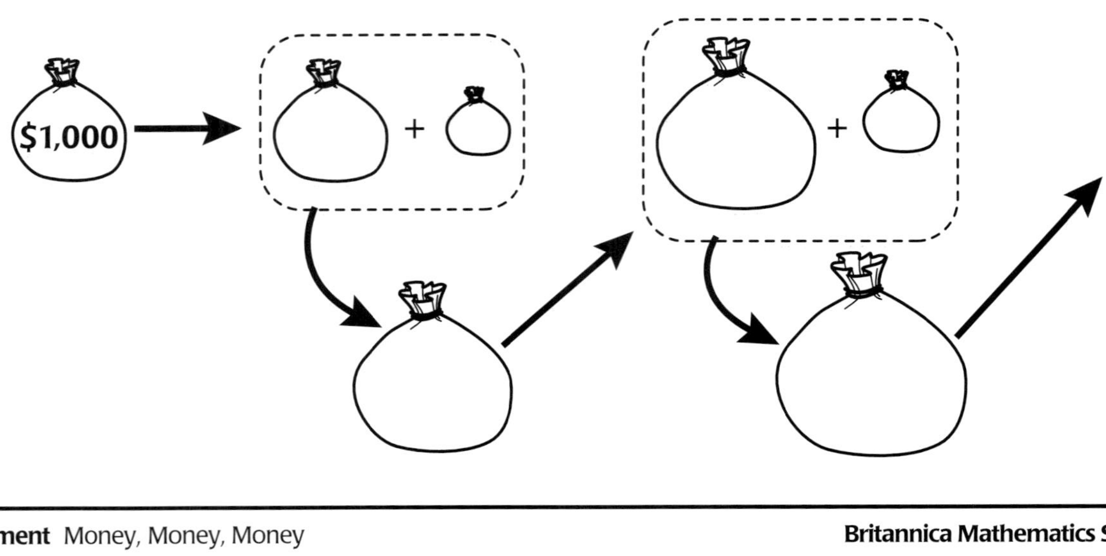

Solutions and Samples
of student work

1. In Jim's daydream, he deposits a certain amount of money in his bank account. The balance increases yearly due to interest on the account.

2.

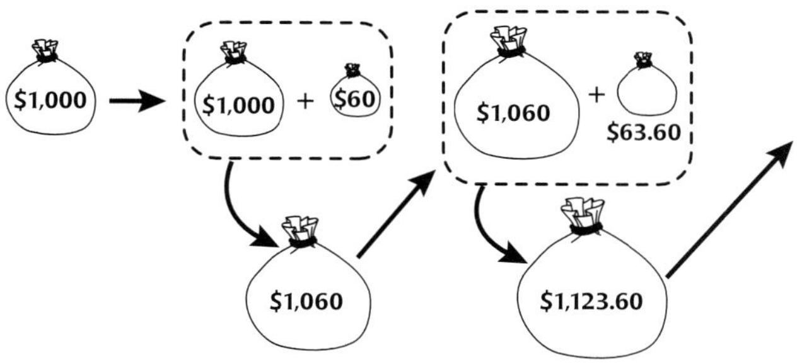

Hints and Comments

Materials Money, Money, Money assessment, pages 128 and 129 of the Teacher Guide (one of each per student); calculators, optional (one per student)

Overview Students investigate to find out how compounded interest affects the balance in a savings account over a two-year period.

About the Mathematics This assessment activity evaluates students' ability to recognize and describe growth patterns presented in pictures, tables, and graphs and their ability to interpret and create direct and recursive formulas for linear and exponential growth.

Planning You may want to have a brief class discussion about the meaning of the term *interest* and the various ways in which bank accounts accrue interest (compounded monthly, quarterly, or yearly). This activity takes approximately one-half of one class session. Students may work individually on these assessment problems.

Comments about the Problems

1. You might mention that most banks compute the quarterly interest on savings accounts based on the account balance of the previous quarter along with the interest accrued during that quarter.

2. Make sure that students understand that they need to take 6% of the amount of money in Jim's bank account at that moment. The annual interest amount grows every year.

MONEY, MONEY, MONEY

Use additional paper as needed.

Jim's bank account balance after the second year can be calculated from his bank account balance during the current year using the recursive formula:

AMOUNT NEXT YEAR = ________ × AMOUNT CURRENT YEAR

3. Fill in the blank in the above formula. What is the growth factor?

4. What will Jim's bank account balance be at the end of the year 2006 if he makes no withdrawals? Please show the computational work you did to determine the answer.

Jim would like to know how many years it will take to double the original amount of money he deposited the first year, but he can't find his calculator. Without using a calculator, it is very time-consuming to determine how fast his money will double.

5. Use a calculator to determine the number of years it will take for Jim to double his money.

A direct formula could be even more helpful than using a calculator.

6. Using the year A.D. 2000 as the start (as year 0), write a direct formula that could be used to determine the number of years needed for Jim to double his money.

7. In problem **5,** you found the number of years needed to double Jim's original deposit. How long will it take for Jim's new account balance to double? Could you answer this question without doing any further calculations? Explain.

3. AMOUNT NEXT YEAR = **1.06** × AMOUNT CURRENT YEAR

   The growth factor is 1.06.

4. Jim's saving account balance will be about $1,418.52. Strategies will vary. Sample strategies:

   $\$1,000 \times 1.06 \times 1.06 \times 1.06 \times 1.06 \times 1.06 \times 1.06 \approx \$1,418.52$

   $\$1,000 \times (1.06)^6 = \$1,000 \times 1.418519 \approx \$1,418.52$

5. Jim's money will double after 12 years. Strategies will vary. Some students may repeatedly multiply the account balance times the interest rate until the account balance doubles. Other students may compute the answer in one step:

   $\$1,000(1.06)^{12} = \$2,012.20$

6. $\$2,000 = \$1,000(1.06)^n$

7. Jim's money will double again after another 12 years. Yes, this question can be answered without any further calculations. Explanations will vary. Some students may reason that any amount of money in a bank account earning interest at this same interest rate (6%) will double in 12 years. Since $(1.06)^{12} \approx 2$, the amount will double in another 12 years.

Materials Money, Money, Money assessment, pages 128 and 129 of the Teacher Guide (one of each per student); calculators, optional (one per student)

Overview Students use recursive and direct formulas to compute the balance of a bank account after several years.

About the Mathematics These assessment problems evaluate students' ability to understand, create, and interpret direct and recursive formulas for exponential growth.

Planning You might suggest that students use the equals sign key on their calculators as a "repeat-operation" feature. Encourage students to work neatly and to carefully explain their work. Students may work individually on these assessment problems.

Comments about the Problems

4–5. Students can use a variety of strategies to solve these problems. Some students may make a table, using the NEXT-CURRENT formula to compute the value for each year. Other students may do repeated calculations on a calculator or use a formula.

6. Students may use symbols or words in their formula.

7. Note the methods students use to solve this problem. Some students may use a guess-and-check strategy, while others may use reasoning.

Many students will not be aware of the fact that any amount of money accruing 6% interest will double in 12 years. In other words, the time needed to double the money in a savings account is independent of the account balance. The doubling is dependent on the base of the exponential function, or the percent interest being paid, rather than on the amount invested. Give only partial credit to students who do not include an adequate explanation.

Did You Know? A rule of thumb that is used to find out how many years it takes for an amount of money to double is often referred to as the "rule of seventy": percent rate × years = 70. So, if the percent rate is known, the number of years can be found by dividing 70 by the percent rate.

STRANGE CREATURES

Use additional paper as needed.

In this assessment activity, you will investigate the growth patterns of three different varieties of fantasy caterpillars. Here is one example of a fantasy caterpillar.

Every fantasy caterpillar has three distinct parts:

H = head
L = segments with legs
S = simple segments

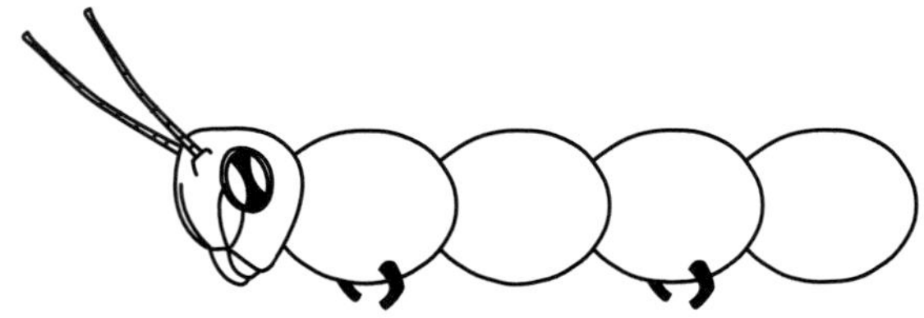

The name of the fantasy caterpillar in the above example is H L S L S, and it has a length of 5.

Variety #1 Fantasy Caterpillars

The growth rules for the variety #1 fantasy caterpillars are as follows:

- The first growth stage begins with a head (H).
- During every subsequent growth stage, one segment with legs (L) and one simple segment (S) are added.

In shorthand, the growth is as follows:

H
↓
H L S

The new segments are added to the right end of each caterpillar. The following drawings show the first three growth stages of a variety #1 fantasy caterpillar.

Stage 1	**H**	Length: 1
	↓	
Stage 2	**H L S**	Length: 3
	↓	
Stage 3	**H L S L S**	Length: 5

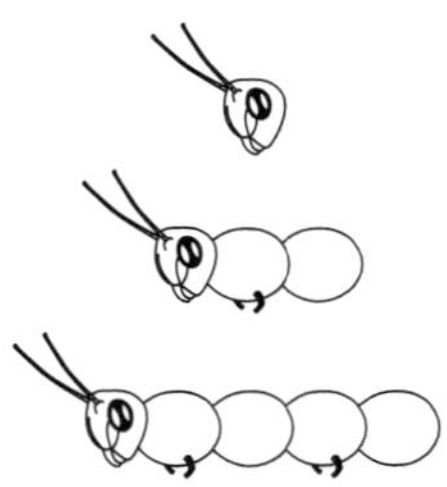

1. Continue the growth pattern of this caterpillar to show stages 4 and 5. You may use drawings or shorthand notation.

2. Determine the length at each stage. Explain the kind of growth that is being exhibited.

3. Predict the length of the fantasy caterpillar at stage 6. Explain.

Solutions and Samples
of student work

1.

Stage 1 **H**  Length: 1

Stage 2 **HLS**  Length: 3

Stage 3 **HLSLS** 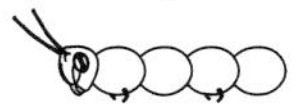 Length: 5

Stage 4 **HLSLSLS** 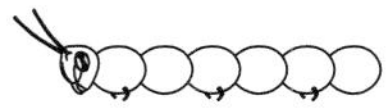 Length: 7

Stage 5 **HLSLSLSLS** 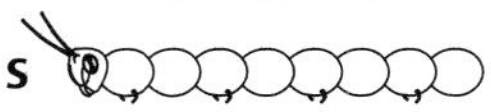 Length: 9

2. See the solution to problem **1** above for the length at each stage. The growth process involved with this variety of caterpillar is linear, or constant. The caterpillar's growth at each stage is determined by adding an L and an S, or a length of 2.

3. The length of the caterpillar at stage 6 will be 11. Some students may express this answer using the shorthand notation HLSLSLSLSLS. Others may make a drawing, as shown below.

HLSLSLSLSLS 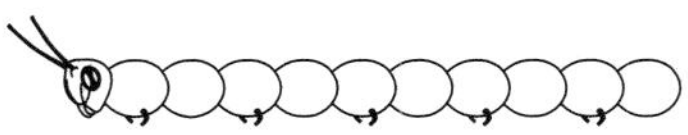

Explanations will vary. Sample explanation:

Since the length of the caterpillar is 9 at stage 5 and a length of 2 is added at each growth stage, the length of the caterpillar at stage 6 will be 11.

Hints and Comments

Materials Strange Creatures assessment, pages 130–132 of the Teacher Guide (one of each per student)

Overview In this assessment activity, students are given growth rules that are used to create pictures of three varieties of fictional caterpillars. Students use the growth rules and given pictures or shorthand notation to determine the lengths of the caterpillars at various stages and to identify the type of growth exhibited by each variety of caterpillar.

About the Mathematics This assessment activity evaluates students' ability to recognize and describe growth patterns presented in pictures, tables, and graphs. It also assesses their ability to create and interpret tables and graphs representing linear growth processes.

Planning Make sure that students understand the symbolic notation used in these assessment problems before they begin answering the questions. This assessment can be completed in one 45-minute class session. Students may work individually on these assessment problems.

Comments about the Problems

1–3. In these problems, students investigate linear growth. Some students may prefer creating and using a direct or recursive formula to determine the length of the caterpillar at each stage, while other students may use the given growth rules to make a drawing of the growth at each stage or use the given symbolic notation to extend the growth pattern.

2. Students should use the word *constant* or *linear* in their explanations and clearly explain the meaning of this type of growth.

3. Students may predict the length at the next growth stage by extending the existing table or by using a formula.

Did You Know? Although the caterpillars in this assessment are fantasy creatures, real caterpillars do grow by adding additional segments during each growth stage.

Use additional paper as needed.

Variety #2 Fantasy Caterpillars

The growth rules for the variety #2 fantasy caterpillars are as follows:

- The first growth stage begins with a head (H).
- During the second growth stage, one segment with legs (L) and one simple segment (S) are added, to make the name of the fantasy caterpillar H L S.
- In each subsequent stage, one L and a number of S's that is one less than the growth stage number are added.

Shorthand notation:

H

↓

H L S

↓

H L S L S S

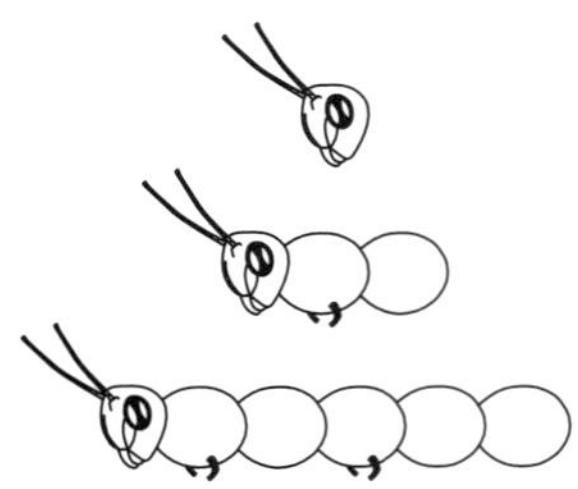

4. Continue the growth pattern of this caterpillar to show stages 4 and 5. You may use drawings or shorthand notation.

5. Determine the length at each stage. Explain the kind of growth that is being exhibited.

6. Predict the length of the fantasy caterpillar in stage 6. Explain.

Solutions and Samples
of student work

4.

Stage 1	**H**
Stage 2	**HLS**
Stage 3	**HLSLSS**
Stage 4	**HLSLSSLSSS**
Stage 5	**HLSLSSLSSSLSSSS**

Stage 1 — Length: 1
Stage 2 — Length: 3
Stage 3 — Length: 6
Stage 4 — Length: 10
Stage 5 — Length: 15

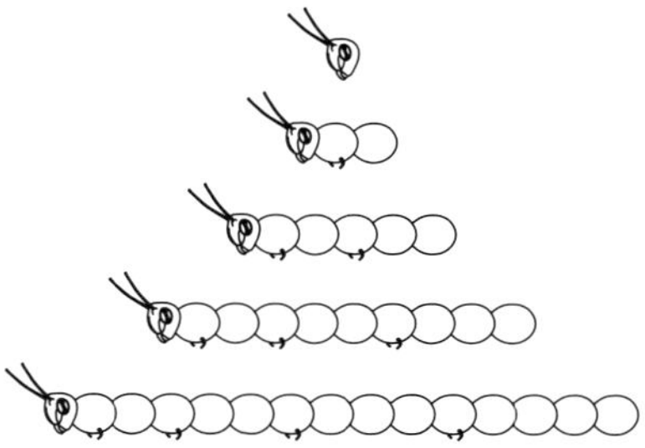

5. (See the solution to problem **4** above for the length at each stage.) Explanations will vary. Students may reason that the growth process involved with this variety of caterpillar is quadratic, or they may simply say that it is increasing. Students may recognize quadratic growth by using a difference table and looking at the second differences to see that they are constant.

Length	1	3	6	10	15
First difference	2	3	4	5	
Second difference	1	1	1		

6. The length of the caterpillar in stage 6 will be 21. Some students may express this answer using shorthand notation:

HLSLSSLSSSLSSSSLSSSSS

Explanations will vary. Sample explanation:

Since the length of the caterpillar is 15 in stage 5 and the length added at each growth stage is one more than that of the previous stage (at stage 2, a length of 2 was added; at stage 3, a length of 3 was added, and so on), the length of the caterpillar in stage 6 will be 21.

Hints and Comments

Materials Strange Creatures assessment, pages 130–132 of the Teacher Guide (one of each per student); calculators, optional (one per student)

Overview Students continue to investigate the growth processes of the caterpillars.

Planning Problems **4–6** and **7–9** are identical to problems **1–3,** although each problem set addresses the growth process of a different type of caterpillar. Encourage students to work systematically and neatly so that they can use the answers to problems **1–9** to answer problem **10.** Students may work individually on these assessment problems.

Comments about the Problems

4–6. In these problems, students investigate quadratic growth. Students may use a formula to determine the length at each growth stage, make drawings to illustrate each growth stage, or extend the growth pattern using symbolic notation.

5. If students are having difficulty identifying this growth process as quadratic, you may suggest that they make and use a difference table. Give only partial credit to such students. They should use the word *quadratic* in their explanations, and clearly explain the meaning of this type of growth.

6. Students may predict the length of the next growth stage by extending the existing table or by using a formula.

Use additional paper as needed.

Variety #3 Fantasy Caterpillars

Here are the growth rules for the variety #3 fantasy caterpillars:

- The first growth stage begins with a head (H).
- During the second growth stage, one segment with legs (L) is added, to make the name of the fantasy caterpillar H L.
- In each subsequent stage are added one L and a number of ending S's that is double the previous number plus one.

Shorthand notation:

H
↓
H L
↓
H L L S
↓
H L L S L S S S

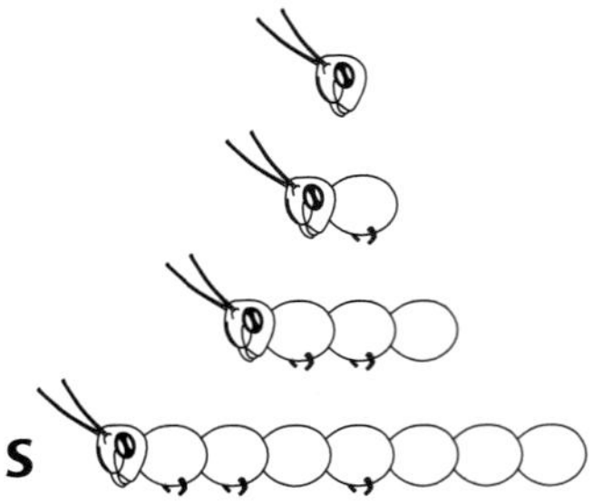

The example above shows the growth for 4 stages.

7. Continue the growth pattern of this caterpillar to show stages 4 and 5. You may use drawings or shorthand notation.

8. Determine the length at each stage. What kind of growth is being exhibited? (You may want to draw an additional stage to get a better impression of the kind of growth.)

9. Predict the length of the fantasy caterpillar in stage 6. Explain.

10. Write a short paragraph that describes the differences in the growth patterns of the three varieties of caterpillars. You may use formulas to describe the three kinds of growth.

Solutions and Samples
of student work

7.

Stage 1	**H**
Stage 2	**H L**
Stage 3	**H L L S**
Stage 4	**H L L S L S S S**
Stage 5	**H L L S L S S S L S S S S S S S**

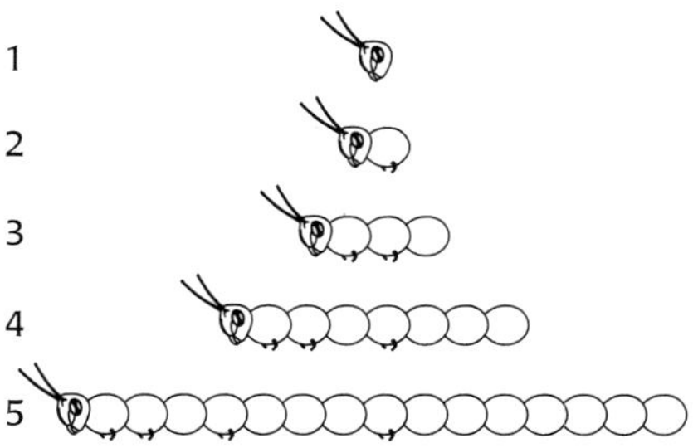

Stage 1 — Length: 1
Stage 2 — Length: 2
Stage 3 — Length: 4
Stage 4 — Length: 8
Stage 5 — Length: 16

8. See the solution to problem **7** above for the length of each stage. Explanations will vary. Students may reason that the growth process involved with this variety of caterpillar is exponential. The next length can be determined by multiplying the current length by two.

9. The length of the caterpillar in stage 6 will be 32. Some students may express this answer using shorthand notation:

HLLSLSSSLSSSSSSSLSSSSSSSSSSSSSSS

10. Answers may vary. Sample response:

The variety #1 caterpillar is growing at a constant rate. The length at each stage can be determined from the previous length by adding an L and an S. The variety #2 caterpillar's growth pattern represents quadratic growth. In quadratic growth, the first differences, the differences between consecutive lengths, increase at a constant rate. The amount of that increase can be determined using a difference table and determining the value of the second difference. The variety #3 caterpillar's growth pattern represents exponential growth. The differences between successive lengths change. In exponential growth, each length is determined by multiplying the previous length by a constant growth factor (in this case, the growth factor is 2).

Hints and Comments

Materials Strange Creatures assessment, pages 130–132 of the Teacher Guide (one of each per student); calculators, optional (one per student)

Overview Students continue to investigate the growth processes of the caterpillars.

About the Mathematics Students' responses to problem 10 will give you insight into their ability to compare three of the different growth processes presented in this unit.

Planning Students may work individually on problems **7–10.**

Comments about the Problems

7–9. In these problems, students investigate exponential (or changing or explosive) growth. Again, students may use a formula to determine the length at each growth stage, make drawings to illustrate each growth stage, or extend the growth pattern using symbolic notation.

8. They should use the word *exponential* or *explosive* in their explanations, and clearly explain the meaning of this type of growth.

9. Again, students may predict the length at the next growth stage by extending the existing table or by using a formula.

10. Students should be able to identify and differentiate between the three growth processes, although at different levels of abstraction.

A GROWTH EXPERIMENT

Use additional paper as needed.

The graph below shows the growth of seven plants (A, B, C, D, E, F, and G) over a six-week period. The plants' heights were measured and recorded weekly.

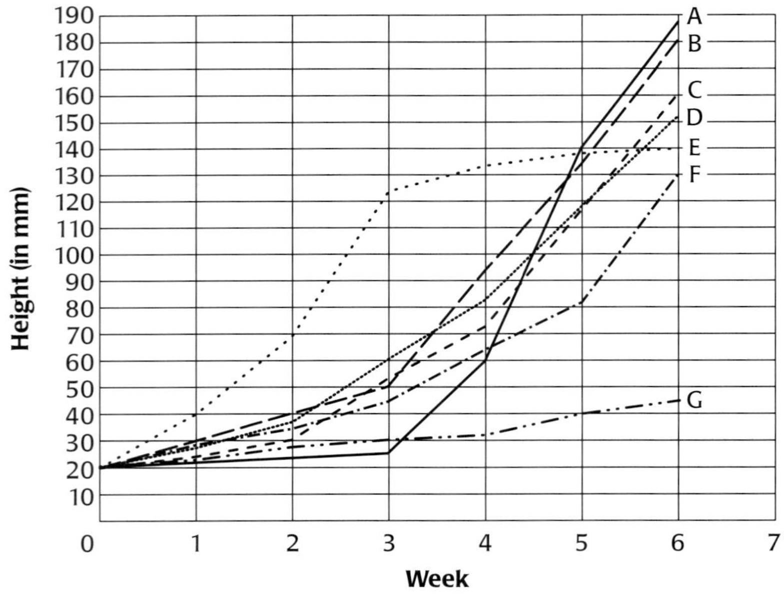

1. Lightly shade a useful band in this figure. Explain how you determined your band.

2. Sketch a curve in the graph that you think shows the typical growth of these plants.

The picture graph below shows the typical growth of these plants for weeks one and two.

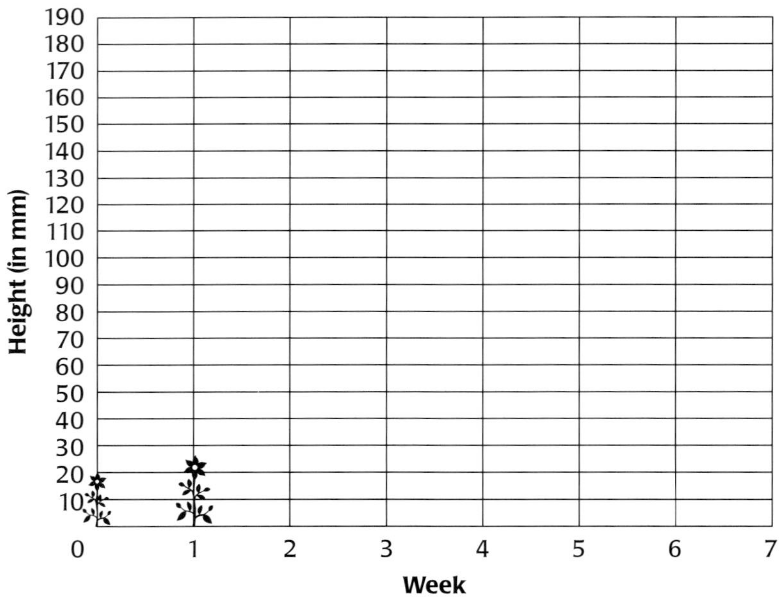

3. Complete the picture graph to show what you think is the typical growth for weeks 2–6. You can use the curve you sketched in problem **2** to help you determine heights.

1. Bands may vary. Sample response:

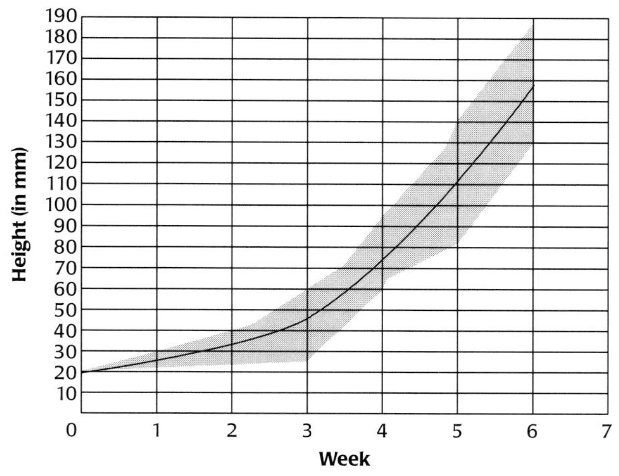

For the band pictured above, the graphs of plants E and G were considered outliers and eliminated.

2. See the curve drawn in the graph for problem **1.**

3.

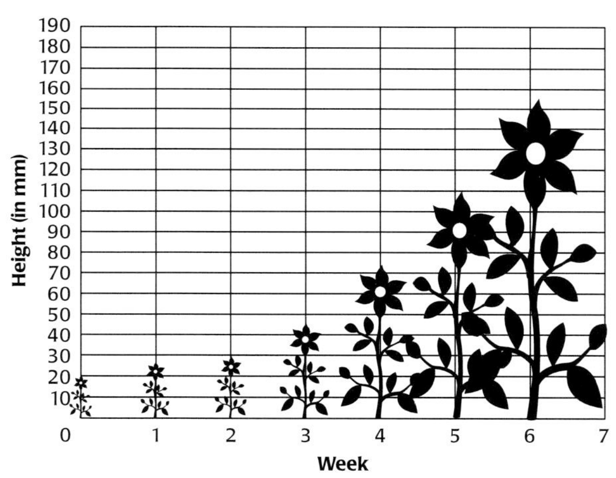

Materials A Growth Experiment assessment, pages 133 and 134 of the Teacher Guide (one of each per student); graphing calculators, optional (one per student); rulers (one per student); colored pencils (three different colors per student)

Overview Students investigate the growth rate of seven plants of the same species, make a band graph, and sketch a curve and draw pictures to represent the typical growth.

About the Mathematics This assessment evaluates students' ability to model a growth situation by choosing an appropriate formula and justify the choice; to recognize that a mathematical model usually does not fit a situation perfectly; and to recognize that growth processes can be simulated with a simplified mathematical model.

Planning Problems **1–3** are time-consuming since they involve detailed graphing and drawing and may require one 45-minute class session. Students may work individually on these assessment problems.

Comments about the Problems

1. Students may use a variety of strategies to come up with an appropriate band. Many band graphs are possible. Focus more on students' explanations than on the accuracy of their band graphs.

2. If students are having difficulty, you might suggest that they draw the curve in the middle of the band graph drawn for problem **1.** Students' curves should be smooth, since growth is usually smooth.

   Many different curves may be drawn to represent the typical growth. As in problem **1,** students' justifications are more important than the accuracy of the curve drawn on the graph.

3. This question allows students to show some creativity. They need to use the curve made in problem **2** to help them make a series of plant drawings that follow the same growth pattern.

A GROWTH EXPERIMENT

Use additional paper as needed.

A suggested formula for the typical growth of these plants is

$$h = 20 \times (1.4)^n$$

where h = the height in millimeters and n = the number of weeks after the first measurement.

4. a. On the diagram from problem **2,** draw the graph of the formula $h = 20 \times (1.4)^n$. Use a different-colored pencil to draw the graph.

b. Draw the same graph of the formula in the picture graph from problem 3.

5. Do you think that this formula realistically depicts the typical growth of these plants? Explain why or why not.

4. a.

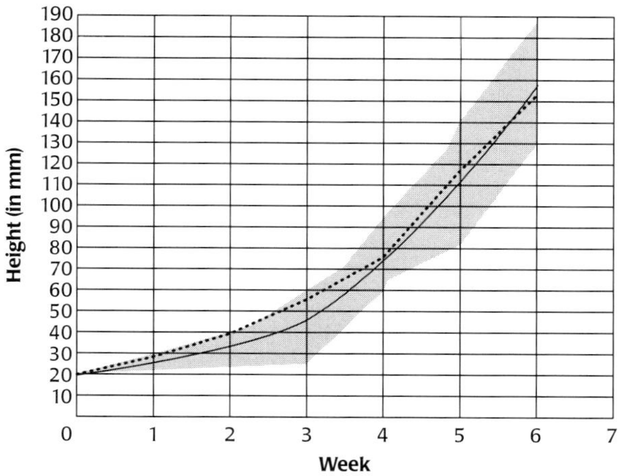

b.

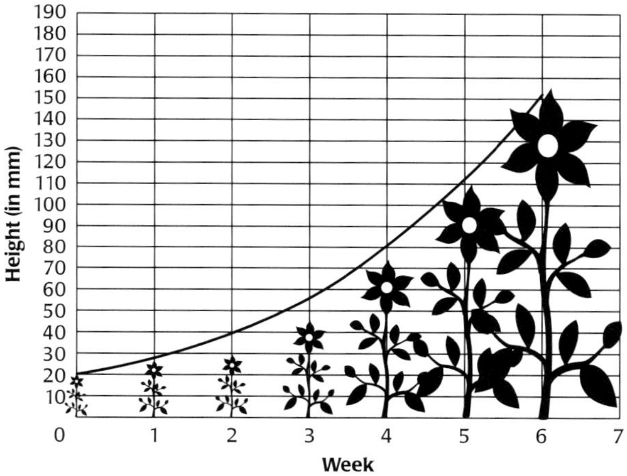

5. Yes, the formula seems to realistically depict the typical growth of these plants. Explanations will vary. Students should note that the plant growth seems to be exponential. Although the plant grows slower in the beginning and faster at the end of the six-week period, a growth factor of 1.4, as shown in the formula, seems to be a close estimate. So, the formula represents the typical growth pattern of these plants.

Materials A Growth Experiment assessment, pages 133 and 134 of the Teacher Guide (one of each per student); graphing calculators, optional (one per student); rulers (one per student); colored pencils (three different colors per student)

Overview Students interpret an exponential formula, draw a graph of the given formula, and judge the appropriateness of the formula as a model for the growth process they have investigated in this assessment.

About the Mathematics These problems assess students' ability to interpret and create tables and graphs representing exponential growth processes and to recognize that a mathematical model usually does not fit a situation perfectly.

Planning Students may work individually on these assessment problems.

Comments about the Problems

4. a. Students should draw the graph of the given formula in the same graph they used for problem **2.** Be sure students use a different color to draw the graph of the formula so that they can compare graphs.

b. Students should draw this graph as a curve in the same picture graph they created for problem **3.**

5. The graph of the formula should look much like the curve drawn in problem **2.**

Extension As an extension to problem **5,** ask students to create a better formula (one that results in a growth process that starts slower and grows faster). They may use a graphing calculator and experiment with different kinds of formulas.

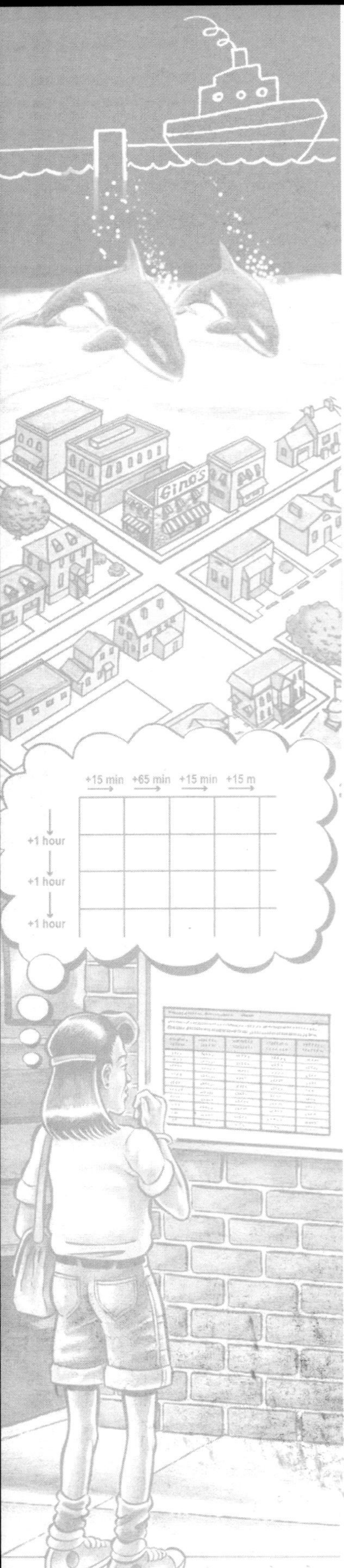

Growth
Glossary

animation (p. 24) simulating a process by using a series of pictures or models (see also simulation)

arrow language (p. 75) a way of describing calculation rules; left of the arrow is the input, above the arrow is the operation, and right of the arrow is the output

base (p. 86) a number that is raised to a power

changing growth (p. 34) growth in which the amount of change between equal time intervals is not constant (see also explosive growth and exponential growth)

constant growth (p. 34) growth in which the amount of change between equal time intervals is constant (see also linear growth)

cubic equation (p. 54) an equation of the form $y = ax^3 + bx^2 + cx + d$

direct formula (p. 22) a formula in which for each independent variable (often x) you get the corresponding value of the dependent variable (often y)

explosive growth (p. 64) growth that increases rapidly (see exponential growth)

exponent (p. 46) the number or variable that indicates how many times to multiply the base by itself; it is written in superscript next to the base

exponential equation (p. 88) an equation in which a variable appears in one of the exponents

exponential growth (p. 88) growth that can be described by an exponential equation

exponential notation (p. 86) writing a number as a power, for instance 10^2.

growth factor (p. 34) the ratio of the amounts of growth for two equal and consecutive time intervals

line graph (p. 8) a graph created by joining consecutive points with a line or a curve

linear growth (p. 26) growth in which the amount of change from one time period to the next is constant; growth that is represented by a linear equation (see also constant growth)

median (p. 19) the middle value in a series of values; if there are two middle values (as there are with an even number of values), the median is the average of these two values

NEXT-CURRENT formula (p. 22) a formula that can be used to find one value when given the previous value

piecewise linear graph (p. 8) a graph created by joining consecutive points with line segments; a method of displaying data using a broken line; a line graph

power (p. 86) a combination of base and exponent

quadratic equation (p. 46) an equation of the form $y = ax^2 + bx + c$

recursive formula (p. 74) see NEXT-CURRENT formula

starting population (p. 68) in a growth process, the value that the process has at $t = 0$

table of differences (p. 42) a table showing the differences between consecutive values; the table of first differences for a quadratic process shows a linear increase

BRITANNICA

Mathematics in Context

Blackline Masters

Letter to the Family

Dear Family,

Very soon your child will begin the *Mathematics in Context* unit *Growth.* Below is a letter to your child that opens the unit, describing the unit and its goals.

In the first section of the unit, your child will study data on the growth of six Wisconsin fast plants over a 17-day period. You and your child might study plant growth at home, too. You could start with several seedlings of the same type of plant. You could then measure the height of each plant every couple of days, and your child could show you how to record the data, graph the data, and describe the growth of the typical plant.

In another section of the unit, your child will use the equation $A = \frac{1}{2}h^2$ to describe the area (A) of a black poplar leaf in terms of its height (h). You and your child could gather several leaves from the same tree and then record the height and area of each leaf. Your child could then show you how to graph the data and find a quadratic equation that describes the area of the leaf in terms of its height. Your child could even show you how to do this on a graphing calculator.

Dear Student,

Welcome to the unit *Growth.*

In this unit, you will study different types of formulas and their graphs in order to determine the best formula for describing the growth of a plant.

Have you ever planted a garden? If so, you have probably noticed that not all plants grow at the same rate. One tomato plant may shoot up right away, but by the end of the next week another tomato plant may overtake it. In this unit, you will study data on several Wisconsin fast plants and then find a way to summarize the data and describe the growth of the typical Wisconsin fast plant.

You will also create models that simulate plant growth, the area of a leaf, and the weight of a snail and find formulas that describe your models. Some of the formulas that you will work with are quadratic equations, such as $y = \frac{1}{2}x^2$, and some are cubic equations, such as $y = 0.07x^3$. You will experiment with different kinds of quadratic and cubic equations on your graphing calculator.

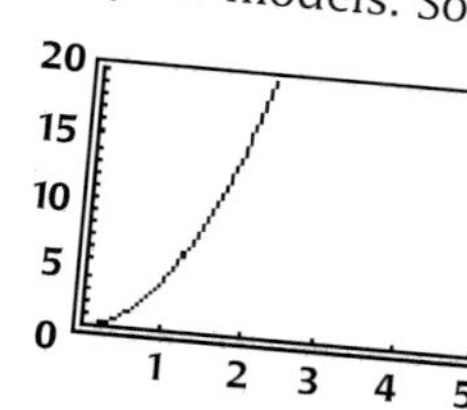

Finally, you will create a model that simulates explosive growth (growth that starts out slowly and then rapidly increases). You will discover that explosive growth is described by an exponential formula, such as $y = 20(0.4)^x$, where 20 is the starting value and 0.4 is the growth factor.

We hope you enjoy modeling these different kinds of growth.

Sincerely,

The Mathematics in Context Development Team

We hope you enjoy exploring different kinds of growth and their corresponding graphs and equations with your child.

Sincerely,

The Mathematics in Context Development Team

Name________________________________

Student Activity Sheet 1

Use with *Growth,* page 7.

14. On Michelle and Erica's graph, sketch a curve that summarizes the data.

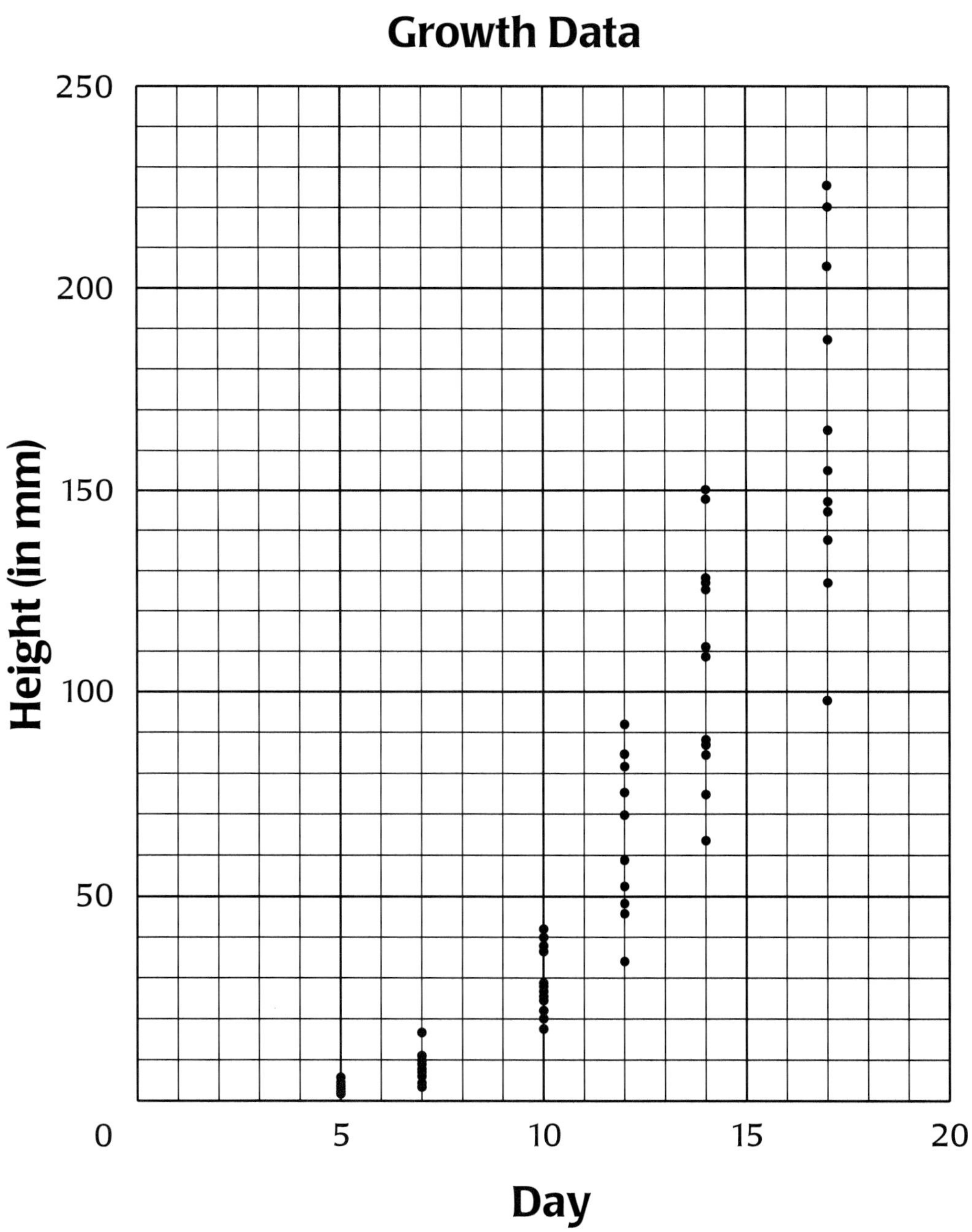

Student Activity Sheet 2

Name ______________________________

Use with *Growth,* page 12.

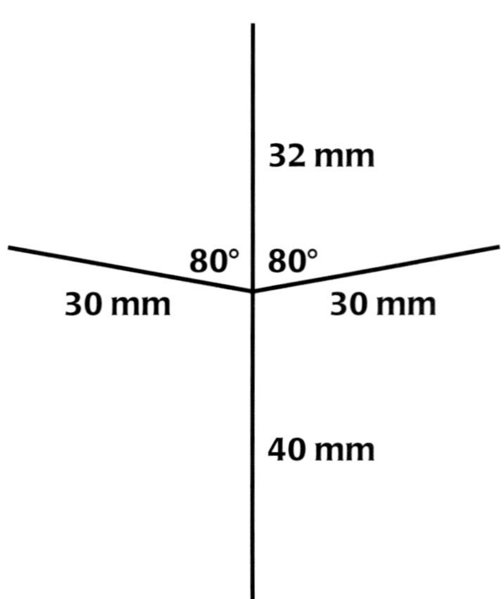

Name____________________

Student Activity Sheet 3

Use with *Growth,* pages 17 and 18.

Height (in cm)	6	7	8	9	10	11	12
Area (in cm^2)	18	24.5					

Height (in cm)	**6**	**7**	**8**	**9**	**10**	**11**	**12**
Area (in cm^2)	**18**	**24.5**	**32**	......	......	......	......
First Difference	**6.5**	**7.5**	......	......	......	......	
Second Difference	**1**	......	......	......	......		

Student Activity Sheet 4

Name ______________________________

Use with *Growth*, page 17.

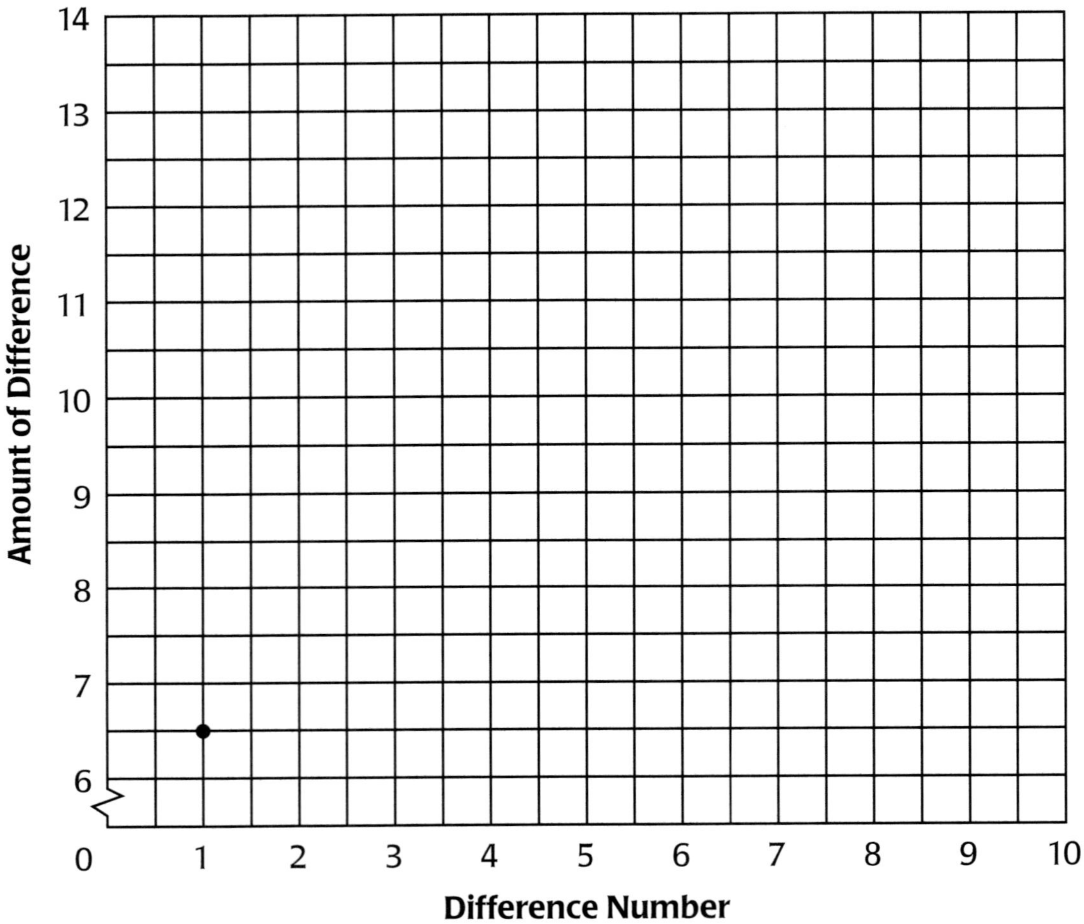

Name____________________________________

Student Activity Sheet 5

Use with *Growth,* page 18.

9. a. Fill in the remaining area values in the table shown below using the formula $A = \frac{1}{2}h^2$ and then graph the formula on the provided grid. Why do you think the graph curves upward?

b. Using your graph, estimate the areas of black poplar leaves with the following heights: 5.5 centimeters, 9.3 centimeters, and 11.7 centimeters.

c. Check your answers to part **b** using the formula for the area of a black poplar leaf. Which method do you prefer for finding the area of a black poplar leaf given its height: the graph or the formula? Explain.

Height (in cm)	1	2	3	4	5	6	7	8
Area (in cm^2)	0.5	2	4.5					

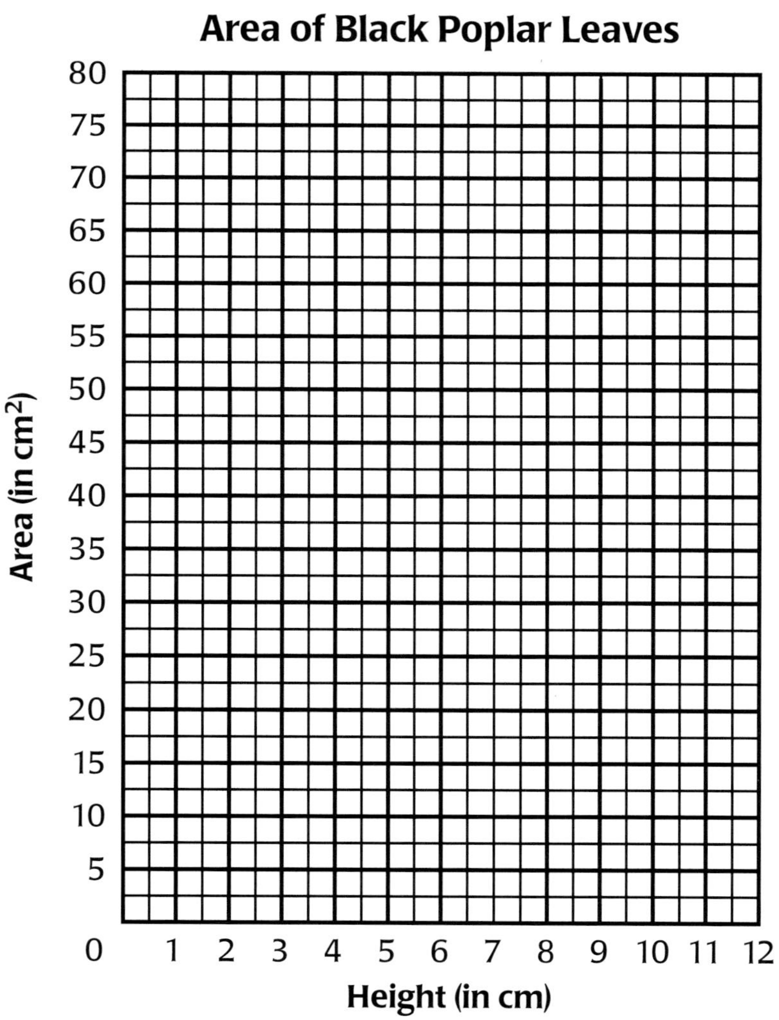

Student Activity Sheet 6

Name ______________________________

Use with *Growth,* page 22.

Making a Scatter Plot

I. Enter the Data

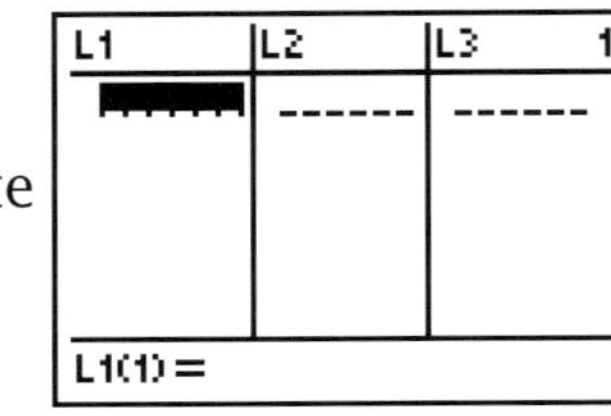

To reach the data tables, press [LIST / STAT] and choose **1:Edit** (press [ENTER] or the number **1**). If there are numbers in lists L1 or L2, you must delete the entries. One way to do this is to use the arrow keys and highlight the list name (L1 or L2) and press [CLEAR] and [ENTER]. Then enter your data for your first list one value at a time under L1, pressing [ENTER] after each value entered. When you have finished, enter your data for your second list under L2.

II. Set up a Viewing Window

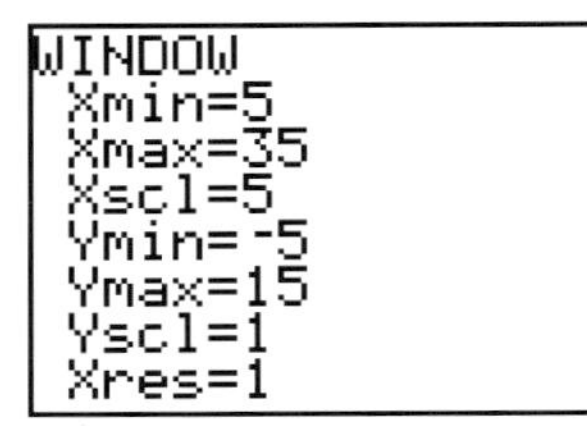

To choose an appropriate viewing window, press [TBLSET / WINDOW]. Then enter the minimum and maximum L1 values for **Xmin** and **Xmax,** and the minimum and maximum L2 values for **Ymin** and **Ymax.** The **Xscl** is the scale along the *x*-axis and **Yscl** is the scale along the *y*-axis. Select appropriate values for each.

III. Make a Scatter Plot

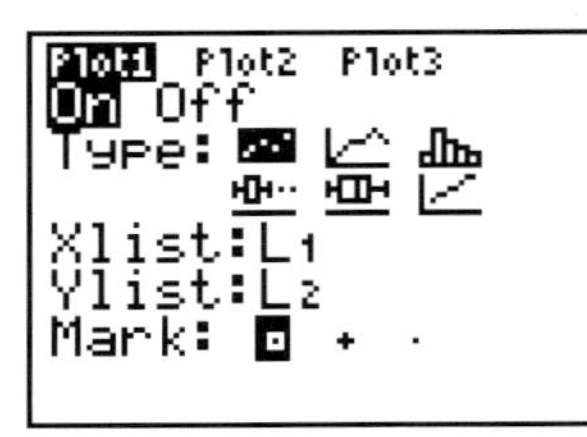

Before you make your scatter plot, make sure that all unwanted plots are off. To choose the scatter plot, press [2nd] [STAT PLOT / Y=] and choose **1:Plot 1.** Then make the selections as shown in the window on the right. When finished, press [2nd] [QUIT / MODE].

Press [TABLE / GRAPH] to see the scatter plot. Use the [CALC / TRACE] key and the left and right arrows to investigate the points.

Name________________________________

Use with *Growth,* page 25.

19. a. On the first grid, graph $y = x^2$ and $y = x^3$ for values of x between 0 and 5 ($0 \leq x \leq 5$). How do the two graphs compare?

b. On the second grid, graph $y = x^2$ and $y = x^3$ for values of x between 0 and 1.5 ($0 \leq x \leq 1.5$). Now how do the two graphs compare?

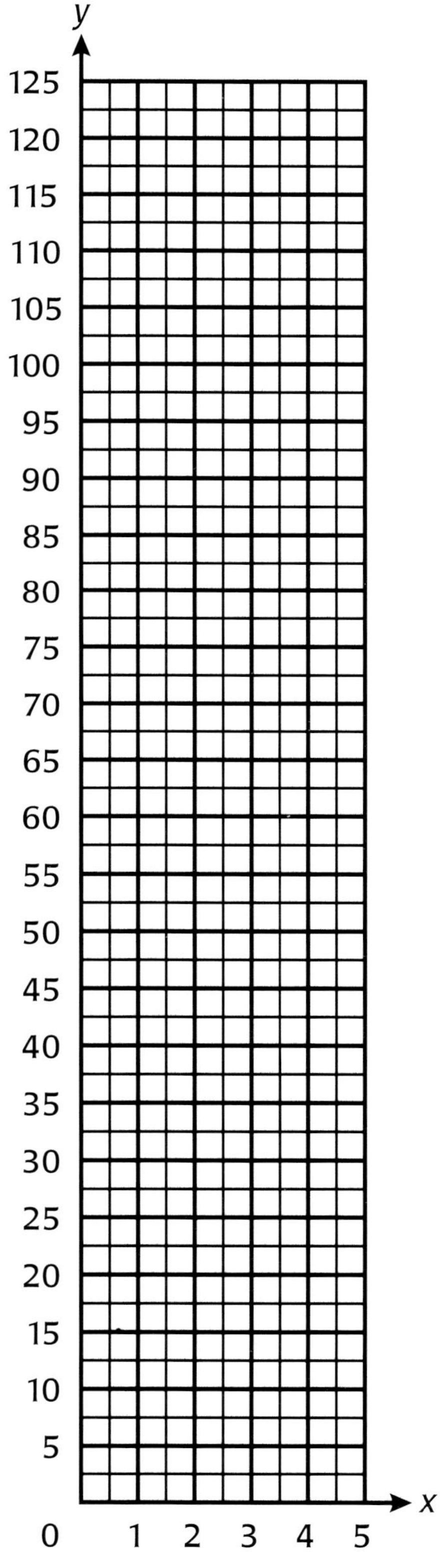

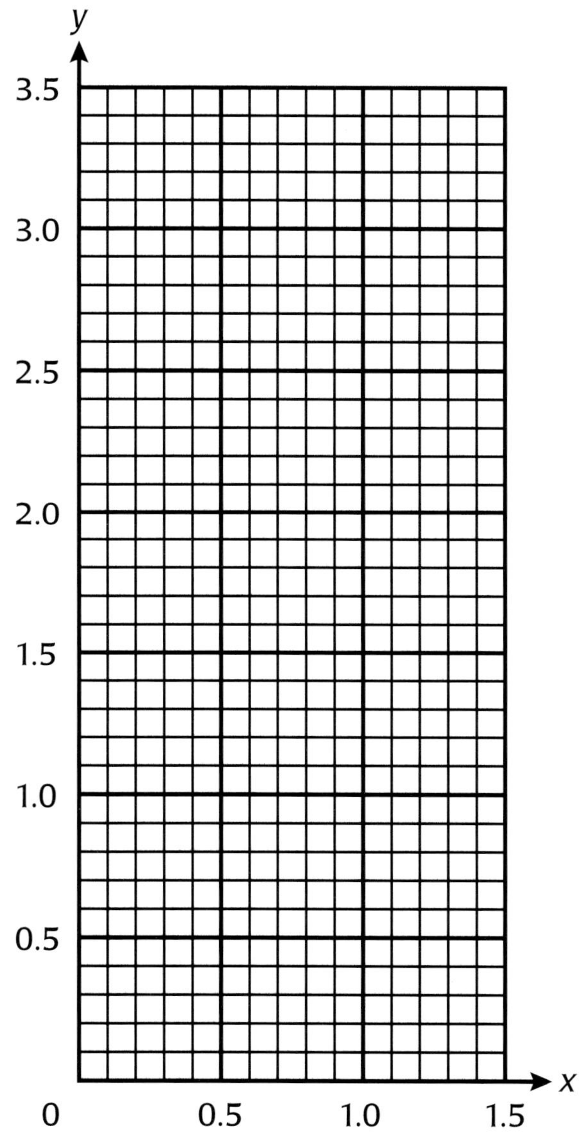

Student Activity Sheet 8

Name ______________________________

Use with *Growth*, pages 29 and 30.

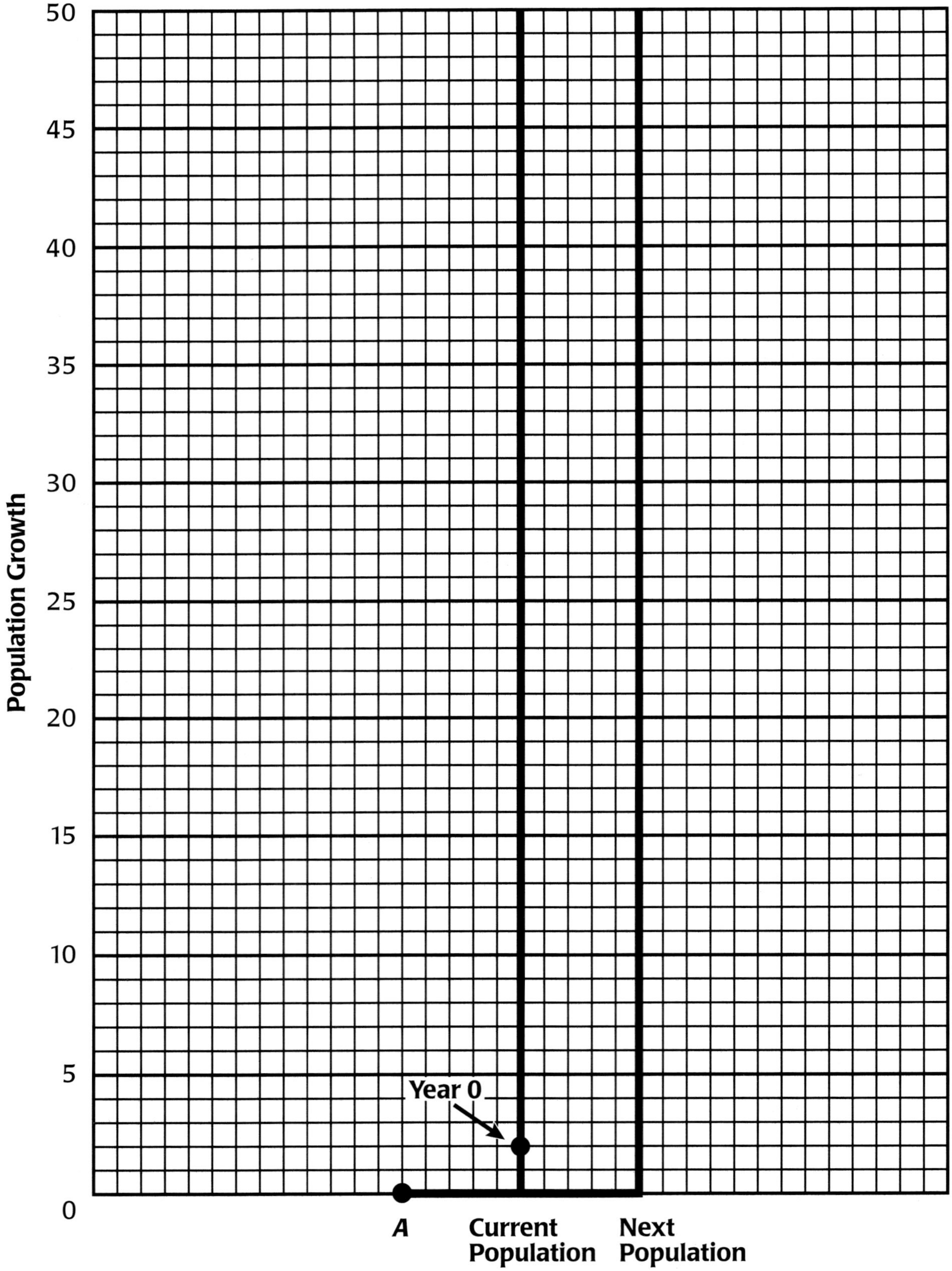

Name________________________________

Student Activity Sheet 9

Use with *Growth*, page 31.

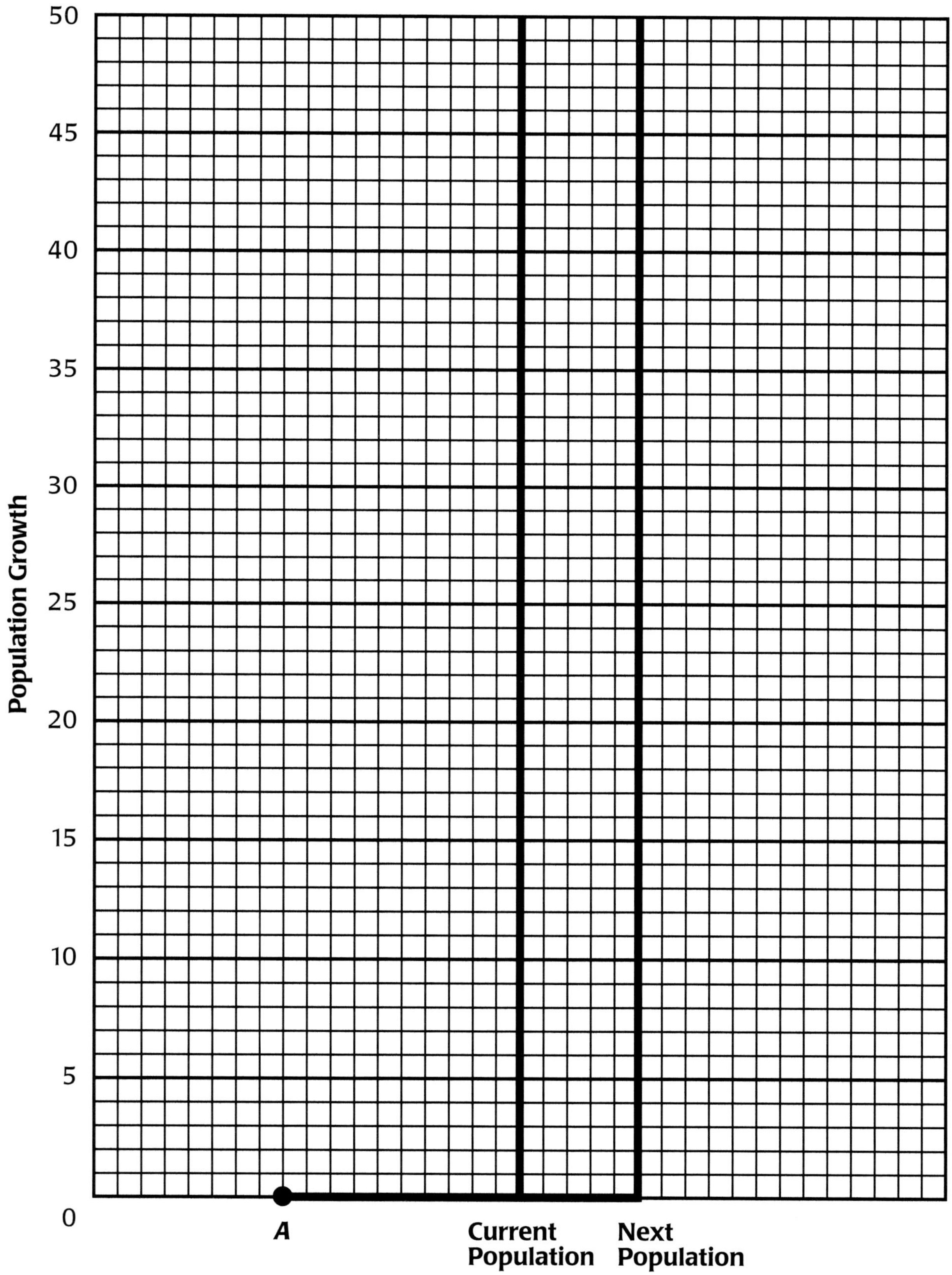

Student Activity Sheet 10

Name ______________________

Use with *Growth*, pages 40 and 41.

Plant E: $h = 20(1.4)^t$

Time (in weeks)	0	1	2	3	4	5	6
Height (in mm)	20						

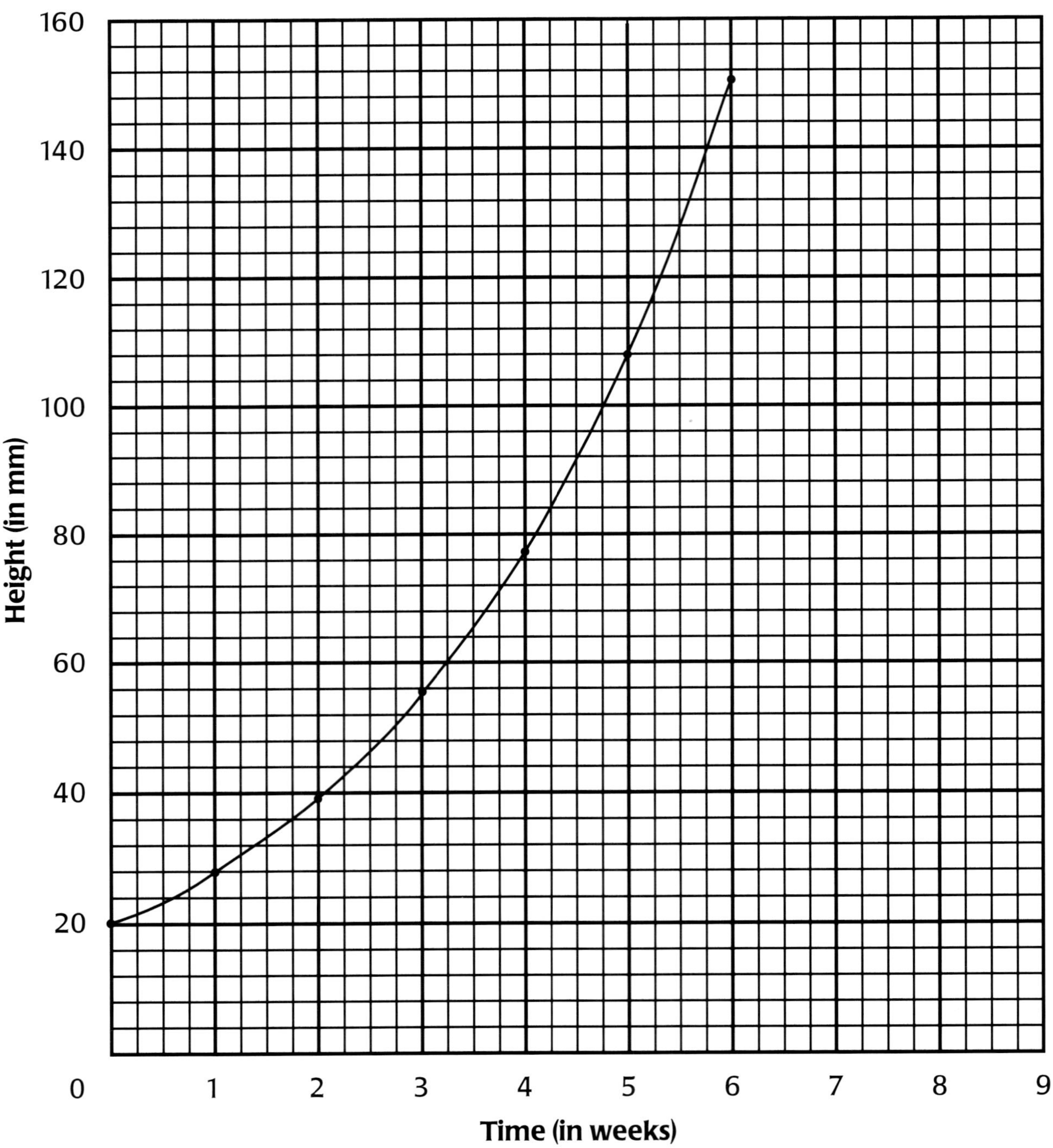

Name____________________________________

Student Activity Sheet 11

Use with *Growth.*

Name ______________________ Date ______________________

MONEY, MONEY, MONEY

Use additional paper as needed.

Jim often daydreams about money. In his last daydream, he deposited a certain amount of money in his bank account and left it there for a two-year period. Here is a picture of Jim's daydream.

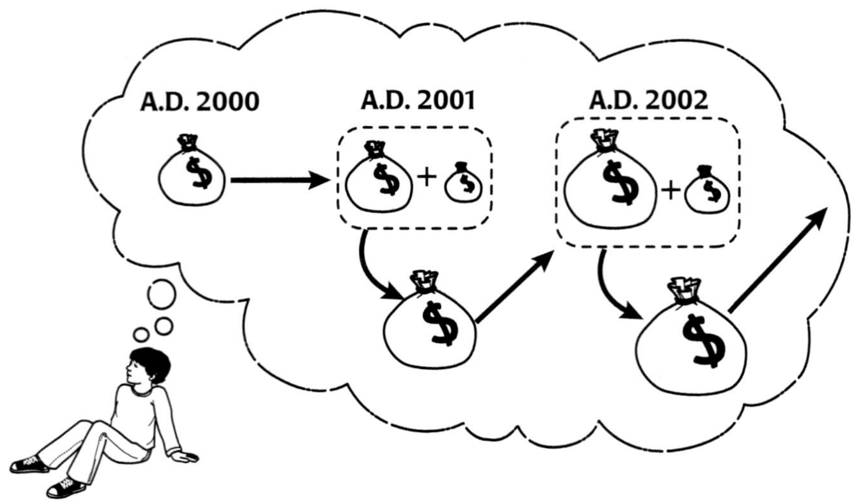

1. Describe what happened in this daydream.

Jim's original bank account deposit in the year A.D. 2000 was $1,000. His bank offers a 6% interest rate compounded annually. This means that when the money has been in the account for a year, Jim gets 6% interest on the sum of the original amount. At the end of the second year, he earns 6% interest on the amount he had at the beginning of that year.

2. Write the correct amount of money on each bag shown in the picture below.

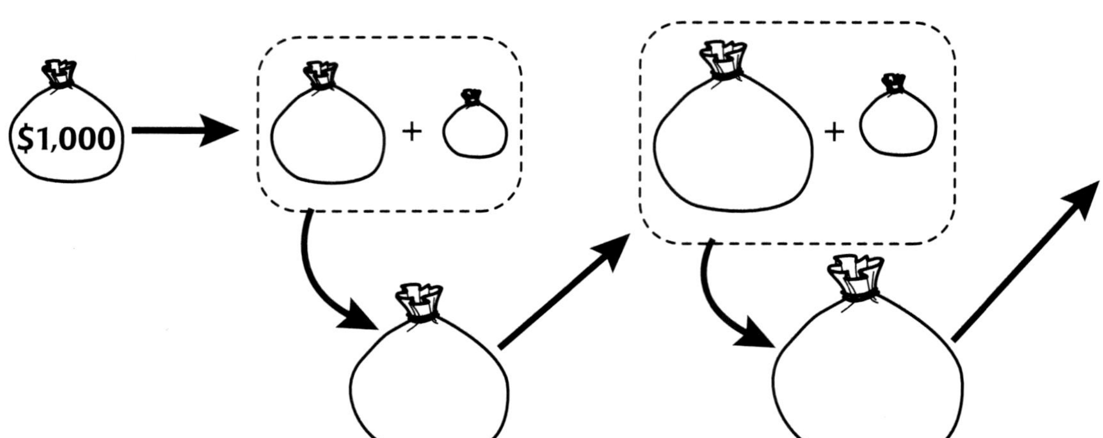

Use additional paper as needed.

Jim's bank account balance after the second year can be calculated from his bank account balance during the current year using the recursive formula:

AMOUNT NEXT YEAR = ________ × AMOUNT CURRENT YEAR

3. Fill in the blank in the above formula. What is the growth factor?

4. What will Jim's bank account balance be at the end of the year 2006 if he makes no withdrawals? Please show the computational work you did to determine the answer.

Jim would like to know how many years it will take to double the original amount of money he deposited the first year, but he can't find his calculator. Without using a calculator, it is very time-consuming to determine how fast his money will double.

5. Use a calculator to determine the number of years it will take for Jim to double his money.

A direct formula could be even more helpful than using a calculator.

6. Using the year A.D. 2000 as the start (as year 0), write a direct formula that could be used to determine the number of years needed for Jim to double his money.

7. In problem **5,** you found the number of years needed to double Jim's original deposit. How long will it take for Jim's new account balance to double? Could you answer this question without doing any further calculations? Explain.

Name ______________________ Date ______________________

STRANGE CREATURES

Use additional paper as needed.

In this assessment activity, you will investigate the growth patterns of three different varieties of fantasy caterpillars. Here is one example of a fantasy caterpillar.

Every fantasy caterpillar has three distinct parts:

H = head
L = segments with legs
S = simple segments

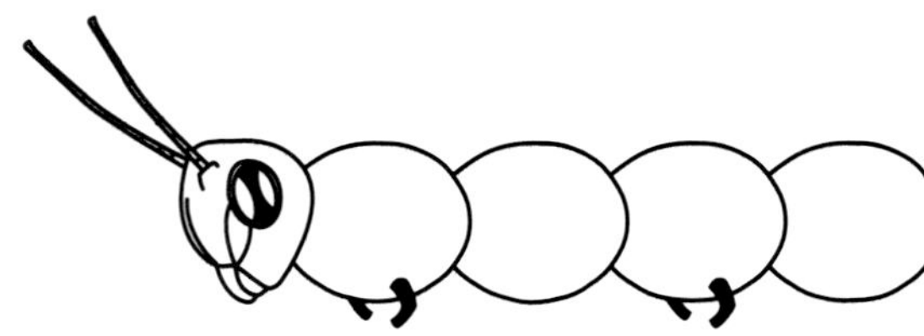

The name of the fantasy caterpillar in the above example is H L S L S, and it has a length of 5.

Variety #1 Fantasy Caterpillars

The growth rules for the variety #1 fantasy caterpillars are as follows:

- The first growth stage begins with a head (H).
- During every subsequent growth stage, one segment with legs (L) and one simple segment (S) are added.

In shorthand, the growth is as follows:

H
↓
H L S

The new segments are added to the right end of each caterpillar. The following drawings show the first three growth stages of a variety #1 fantasy caterpillar.

Stage 1	**H** ↓	Length: 1
Stage 2	**H L S** ↓	Length: 3
Stage 3	**H L S L S**	Length: 5

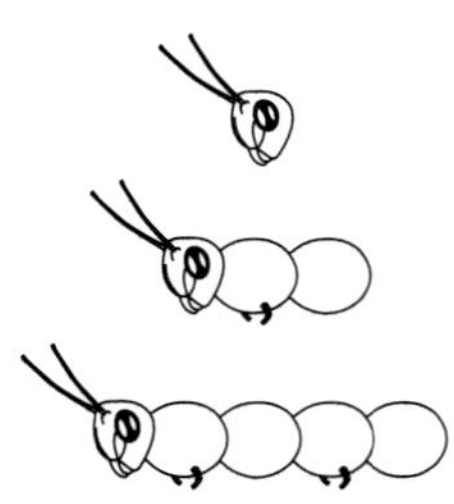

1. Continue the growth pattern of this caterpillar to show stages 4 and 5. You may use drawings or shorthand notation.

2. Determine the length at each stage. Explain the kind of growth that is being exhibited.

3. Predict the length of the fantasy caterpillar at stage 6. Explain.

Name ______________________ Date ______________________

Use additional paper as needed.

Variety #2 Fantasy Caterpillars

The growth rules for the variety #2 fantasy caterpillars are as follows:

- The first growth stage begins with a head (H).
- During the second growth stage, one segment with legs (L) and one simple segment (S) are added, to make the name of the fantasy caterpillar H L S.
- In each subsequent stage, one L and a number of S's that is one less than the growth stage number are added.

Shorthand notation:

H

↓

H L S

↓

H L S L S S

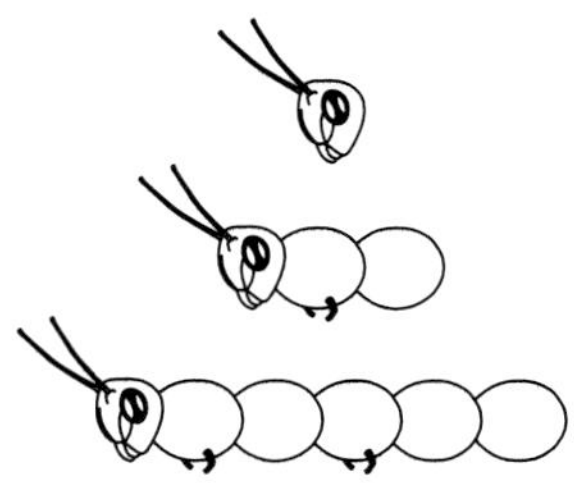

4. Continue the growth pattern of this caterpillar to show stages 4 and 5. You may use drawings or shorthand notation.

5. Determine the length at each stage. Explain the kind of growth that is being exhibited.

6. Predict the length of the fantasy caterpillar in stage 6. Explain.

Name ______________________ Date ______________________

STRANGE CREATURES

Use additional paper as needed.

Variety #3 Fantasy Caterpillars

Here are the growth rules for the variety #3 fantasy caterpillars:

- The first growth stage begins with a head (H).
- During the second growth stage, one segment with legs (L) is added, to make the name of the fantasy caterpillar H L.
- In each subsequent stage are added one L and a number of ending S's that is double the previous number plus one.

Shorthand notation:

H
↓
H L
↓
H L L S
↓
H L L S L S S S

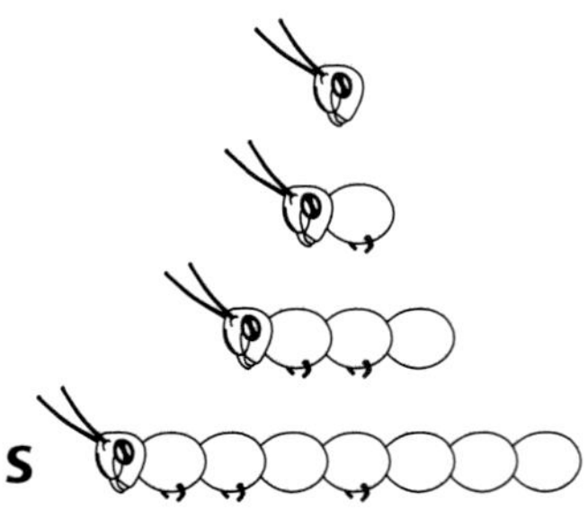

The example above shows the growth for 4 stages.

7. Continue the growth pattern of this caterpillar to show stages 4 and 5. You may use drawings or shorthand notation.

8. Determine the length at each stage. What kind of growth is being exhibited? (You may want to draw an additional stage to get a better impression of the kind of growth.)

9. Predict the length of the fantasy caterpillar in stage 6. Explain.

10. Write a short paragraph that describes the differences in the growth patterns of the three varieties of caterpillars. You may use formulas to describe the three kinds of growth.

Name ______________________ Date ______________________

A GROWTH EXPERIMENT

Use additional paper as needed.

The graph below shows the growth of seven plants (A, B, C, D, E, F, and G) over a six-week period. The plants' heights were measured and recorded weekly.

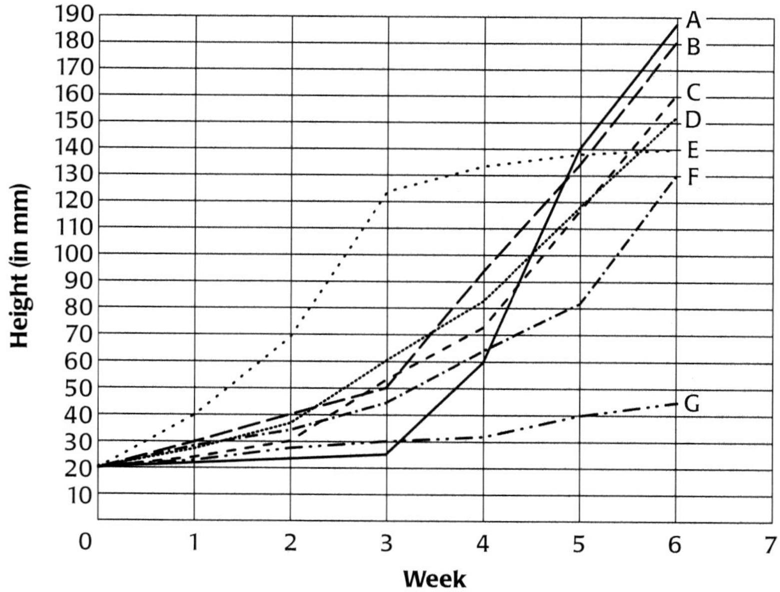

1. Lightly shade a useful band in this figure. Explain how you determined your band.

2. Sketch a curve in the graph that you think shows the typical growth of these plants.

The picture graph below shows the typical growth of these plants for weeks one and two.

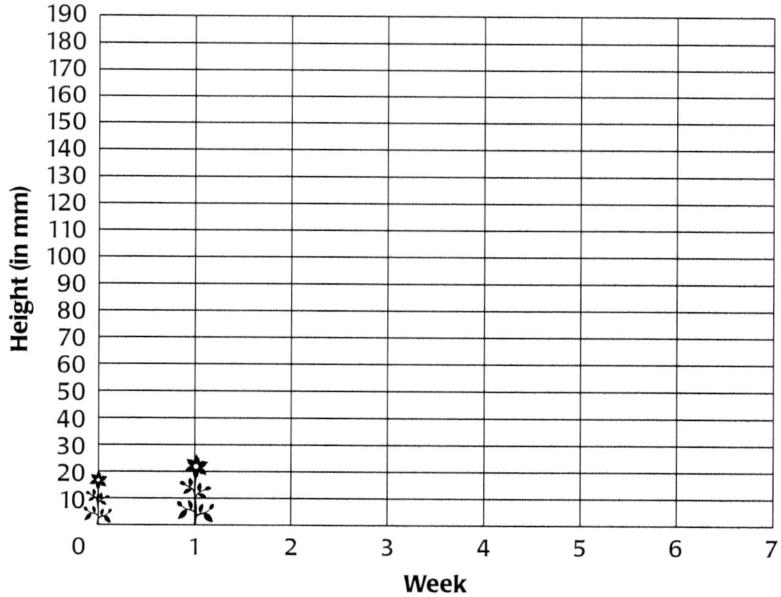

3. Complete the picture graph to show what you think is the typical growth for weeks 2–6. You can use the curve you sketched in problem **2** to help you determine heights.

Name ______________________ Date ______________________

A GROWTH EXPERIMENT

Use additional paper as needed.

A suggested formula for the typical growth of these plants is

$$h = 20 \times (1.4)^n$$

where h = the height in millimeters and n = the number of weeks after the first measurement.

4. a. On the diagram from problem **2,** draw the graph of the formula $h = 20 \times (1.4)^n$. Use a different-colored pencil to draw the graph.

b. Draw the same graph of the formula in the picture graph from problem 3.

5. Do you think that this formula realistically depicts the typical growth of these plants? Explain why or why not.

Section A. Plant Growth

1–2a.

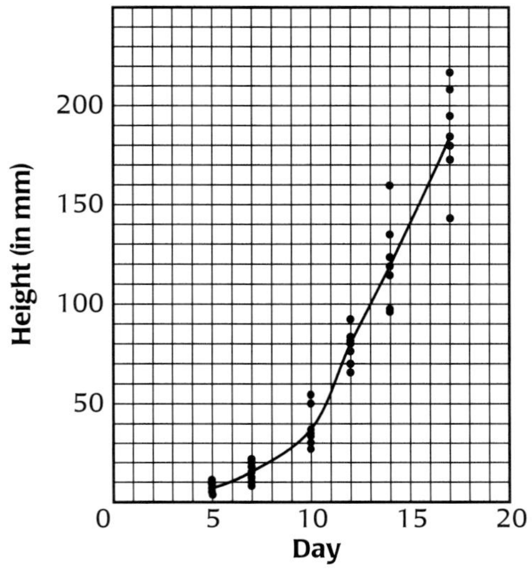

2. b. Answers will vary. Sample explanation:

The median growth graph shows the typical growth of a plant, but it doesn't show the variation among the plants. For example, if one particular plant did not grow at all from day 1 to day 6, this would not show up on the median growth graph.

Section B. Animation and Simulation

1.

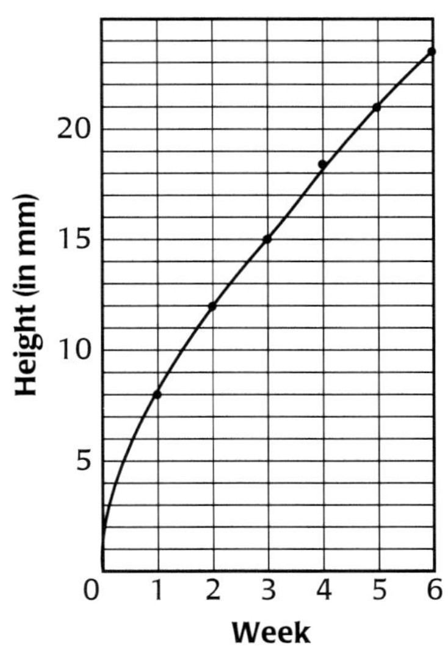

2. No, the constant growth model does not apply. The plant does not grow constantly; in the beginning, the growth is faster than at the end.

3. The growth factor is 0.75.

Section B. (continued)

4.

Week	1	2	3	4	5	6
Length of New Stem Part (in cm)	8	6	4.5	3.375	2.53	1.9

5. The changing growth formula fits better, since the differences between the heights are not constant.

Section C. Equations for Growth

1. B (A cube has six faces, each with an area of x^2.)

2. D (Half of x^3 equals $0.5x^3$.)

3. E (A cube with sides twice as long is a cube that can hold eight of the original cubes; each side now becomes $2x$, so the volume will be $(2x)^3$, or 2^3x^3, which is $8x^3$.)

4. A (A square with sides of double length can be covered by four original squares.)

5. C (Two cubes have a total of 12 faces, but two of these faces are not on the surface.)

Section D. Explosive Growth

1. **a.** The growth factor is 1.5.

   b. The growth factor is 1.5.

2. The growth factor gets smaller.

3. Point *A* should be about 3 units to the left of the Current Population line. Students will find this point by drawing lines through applicable current and next population points.

4. Here is the growth diagram with a starting population of 1.

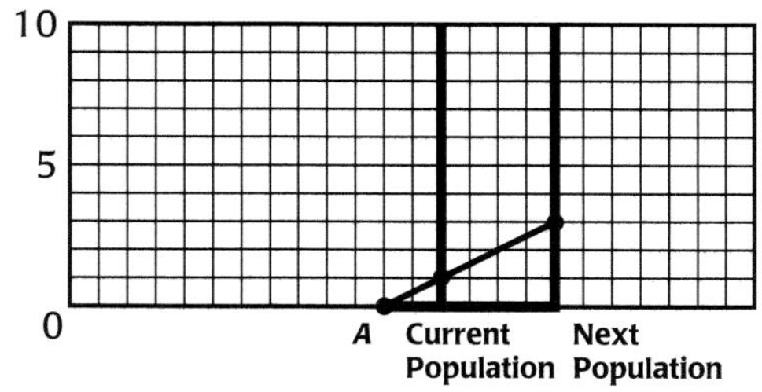

Starting Population 1

Year	Population
0	1
1	3
2	9
3	27

Here are two more examples:

Starting Population 2

Year	Population
0	2
1	6
2	18
3	54

Starting Population 3

Year	Population
0	3
1	9
2	27
3	81

Section E. Exponential Growth

1.

Time (in hours)	1	2	3	4	5	6
Medicine Left in Stomach	0.5	0.25	0.125	≈ 0.06	≈ 0.03	≈ 0.015

2. NEXT AMOUNT = CURRENT AMOUNT × 0.5

   $M = 1(0.5)^t$, where *M* is the amount of medicine left in the stomach and *t* is the number of hours that have passed.

3. **a.** 18.75 milligrams $[600(0.5)^5]$

   b. $M = 600(0.5)^t$

4. **a.** *M* stands for the amount of medicine remaining in Sam's stomach; 450 is the number of milligrams of medicine he took; 0.6 is the growth factor that tells how much of the previous amount is left after each time period; and *t* is the number of hours that have gone by.

   b. about 35 milligrams $[450(0.6)^5 \approx 35]$

CREDITS

Cover

Design by Ralph Paquet/Encyclopædia Britannica Educational Corporation.

Collage by Koorosh Jamalpur/KJ Graphics.

Title Page

Phil Geib/Encyclopædia Britannica Educational Corporation.

Illustrations

6, 16, 18, 20, 28, 38, 40, 42, 46, 48, 50, 52, 54, 68, 76, 88, 90, 94 Phil Geib/Encyclopædia Britannica Educational Corporation.

Photographs

10, 12 © Linda Baham/*Wisconsin Fast Plants*, University of Wisconsin–Madison, Department of Plant Pathology; **56** © James H. Carmichael Jr./The Image Bank; **64** © James P. Rowan/Tony Stone Images; **66** © Ron Sanford/Tony Stone Images.